Architecting Vue.js 3 Enterprise-Ready Web Applications

Build and deliver scalable and high-performance, enterprise-ready applications with Vue and JavaScript

Solomon Eseme

BIRMINGHAM—MUMBAI

Architecting Vue.js 3 Enterprise-Ready Web Applications

Group Product Manager: Pavan Ramchandani

Publishing Product Manager: Aaron Tanna

Senior Content Development Editor: Rakhi Patel

Technical Editor: Saurabh Kadave

Copy Editor: Safis Editing

Project Coordinator: Manthan Patel

Proofreader: Safis Editing

Indexer: Sejal Dsilva

Production Designer: Prashant Ghare

Marketing Coordinators: Namita Velgekar, Nivedita Pandey, and Anamika Singh

First published: April 2023

Production reference: 1170323

Published by Packt Publishing Ltd.

Livery Place

35 Livery Street

Birmingham

B3 2PB, UK.

ISBN 978-1-80107-390-5

www.packtpub.com

To my parents Mr/Mrs. Eseme Adaada, family, and friends, for their sacrifices and for exemplifying the power of determination and resilience.

– Solomon Eseme

Contributors

About the author

Solomon Eseme is an experienced software engineer, content creator, and the founder of Mastering Backend, with 5+ years of experience working across multiple frontend and backend technologies to design and build high-performing, scalable, and innovative applications, following best practices and industry standards in a variety of workplaces, from start-ups to larger consultancies. He started using Vue when it was first integrated with Laravel and has never looked back. He is also a panelist on the ViewsOnVue podcast and a technical writer with Vue.js developers.

I want to thank the people who have been close to me and supported me, especially my parents, family, and close friends, Success Ibekwe, Moses Anumadu, and Godknows Eseme(Boikay) (brother).

.

About the reviewer

Brice Chaponneau is a freelance solution architect and also offers his services as an auditor or lead in current frameworks, such as Vue, React, Svelte, and Solid. He has worked in various domains, with clients such as Arcelor Mittal, Caisse d'Epargne, SNCF, Société Générale, Natixis, Edmond de Rothschild, Carrefour, Galeries Lafayette, KPMG, and Louis Vuitton. He has written a book on Vue 2, published by Eyrolles in 2019. He participated as a reviewer on the book *Building Vue.js Application with GraphQL* for Packt in 2020.

Table of Contents

Part 2: Large-Scale Apps and Scaling Performance in Vue.js 3

3

Scaling Performance in Vue.js 3   33

4

Architecture for Large-Scale Web Apps   45

Part 3: Vue.js 3 Enterprise Tools

5

6

7

Part 4: Testing Enterprise Vue.js 3 Apps

8

Testing and What to Test in Vue.js 3 — 127

9

Best Practices in Unit Testing — 145

10

Integration Testing in Vue.js 3 159

11

Industry-Standard End-to-End Testing 173

Part 5: Deploying Enterprise-ready Vue.js 3

12

Deploying Enterprise-Ready Vue.js 3 189

Preface

Building enterprise-ready Vue.js apps entails following best practices to create high-performance and scalable Vue.js applications.

This book is a must for any developer who works with a large Vue.js code base where performance and scalability are important. You'll learn how to configure and set up Vue.js 3 and Composition API, and how to build real-world applications with it. You'll then learn to create reusable components in Vue.js 3 and scale performance in Vue.js 3 applications.

You will then learn to scale performance with asynchronous lazy loading, image compression, code splitting, and tree shaking. Next, you'll learn how to use RESTful API, Docker, GraphQL, and different types of testing to ensure that your Vue.js 3 application is scalable and maintainable. By the end of this book, you will be able to build and deploy your enterprise-ready Vue.js 3 application using best practices in implementing RESTful API, Docker, GraphQL, and different testing methods with Vue.js 3.

Who this book is for

The book is intended for Vue.js developers and professional frontend developers who want to build high-performance, production-grade, and enterprise-scalable Vue.js apps from design to deployment. The book assumes a working knowledge of Vue.js and JavaScript programming.

What this book covers

Chapter 1, *Getting Started with Vue.js 3*, covers Vue.js, Options API, the new Vue.js 3, and Composition API. Thus, it establishes and fosters an understanding of Vue.js. In addition, this chapter will explore the Vue.js 3 Composition API in depth and serve as a guide to understanding the other chapters.

Chapter 2, *Using Libraries for Large-Scale Application*, covers the essential aspects of Vuex, Axios, and Vue Router and how to integrate them with Vue 3 to develop an enterprise application. This background information will put you in a better position to grasp the terms and concepts of these libraries and help you understand how to build and scale an enterprise-ready application.

Chapter 3, Scaling Performance in Vue.js 3, dives deeper into scaling an extensive Vue application. You will learn how to scale performance with asynchronous lazy loading, image compression, code splitting, tree shaking, and many other tricks to increase the performance of your Vue.js 3 enterprise-ready application.

Chapter 4, Architecture for Large-Scale Web Applications, teaches you how to handle a sizable enterprise-ready project, from managing larger file structures to using the micro-frontend architecture. You will also learn how to handle the internationalization and localization of your Vue.js 3 project.

Chapter 5, An Introduction to GraphQL, Queries, Mutations, and RESTful APIs, explores GraphQL, Apollo Server 2, queries, mutations, and how to integrate these technologies into your Vue.js 3 application. In addition, you will learn how to utilize GraphQL to deliver scalable and high-performance applications.

Chapter 6, Building a Complete Pinterest Clone with GraphQL, discusses how to build a complete Pinterest clone with Vue 3 and GraphQL. You will utilize the knowledge of GraphQL to develop and deliver an enterprise application, such as Pinterest, using Vue 3 and GraphQL.

Chapter 7, Dockerizing a Vue 3 App, examines the nitty-gritty involved in dockerizing your Vue project. In addition, you will learn the best practices and industry standards to dockerize and deploy an enterprise Vue.js 3 web application. This chapter will also go more practical by dockerizing a full-stack web application and deploying a container to a cloud platform using Docker Compose. You will learn how to handle larger projects with this tool.

Chapter 8, Testing and What to Test in Vue.js 3, explores the whole concept of testing. You will learn what to test from an available array of components and methods. In addition, you will learn the best practices and industry standards for testing libraries and how to integrate them with Vue.js 3.

Chapter 9, Best Practices in Unit Testing, dives into everything related to unit testing. You will learn how to unit-test a Vue.js 3 component and the components and pages' methods. You will also learn unit tools such as Jest and Mocha and use them to effectively unit-test an enterprise project.

Chapter 10, Integration Testing in Vue.js 3, covers everything related to integration testing. You will learn in depth how to perform an integration test on a Vue.js 3 component and pages. You will also learn about integration testing tools such as Vue-Test-Library and how to use them to test an enterprise project effectively.

Chapter 11, Industry Standard End-to-End Testing, explores everything related to end-to-end testing. You will learn in depth how to perform end-to-end testing on a Vue.js 3 component and pages. In addition, you will also learn about end-to-end testing tools, such as Cypress and Puppeteer, and how to use them to test an enterprise project end to end effectively.

Chapter 12, Deploying Enterprise-Ready Vue.js 3, shows you how to deploy Vue.js 3 projects to the AWS cloud. You will learn the best practices in deploying to AWS. In addition, you will learn how big companies deploy their enterprise Vue applications.

Chapter 13, *Advanced Vue.js Frameworks*, offers a definitive guide to Nuxt.js. You will learn the nitty-gritty of Nuxt.js and how to build and deliver enterprise SSR projects with Vue.js 3. We will also offer a definitive guide to Gridsome. You will learn the nitty-gritty of Gridsome and how to build and deliver enterprise CSR projects with Vue.js 3.

To get the most out of this book

Software/hardware covered in the book	Operating system requirements
Node.js 16.0 or higher	Windows, macOS, or Linux
Familiarity with command line	Windows, macOS, or Linux
JavaScript, ECMAScript 11	Windows, macOS, or Linux
Vue.js 3	Windows, macOS, or Linux
Docker and AWS	Windows, macOS, or Linux

If you are using the digital version of this book, we advise you to type the code yourself or access the code from the book's GitHub repository (a link is available in the next section). Doing so will help you avoid any potential errors related to the copying and pasting of code.

You will need knowledge of Docker and containerization to run the code presented in Chapter 7.

You will need knowledge of AWS and cloud computing to run the code presented in Chapter 12.

Download the example code files

You can download the example code files for this book from GitHub at `https://github.com/PacktPublishing/Architecting-Vue.js-3-Enterprise-Ready-Web-Applications`. If there's an update to the code, it will be updated in the GitHub repository.

We also have other code bundles from our rich catalog of books and videos available at `https://github.com/PacktPublishing/`. Check them out!

Download the color images

We also provide a PDF file that has color images of the screenshots and diagrams used in this book. You can download it here: `https://packt.link/4Lgta`.

Conventions used

There are a number of text conventions used throughout this book.

Code in text: Indicates code words in text, database table names, folder names, filenames, file extensions, pathnames, dummy URLs, user input, and Twitter handles. Here is an example: "If you're following along, create a new file called staging.yml inside the .github/workflows folder."

A block of code is set as follows:

```
lint:
  runs-on: ubuntu-latest
  steps:
    - uses: actions/checkout@v3
    - run: |
        yarn
        yarn lint
```

Any command-line input or output is written as follows:

```
npm install --save graphql graphql-tag @apollo/client @vue/
apollo-composable
```

Bold: Indicates a new term, an important word, or words that you see on screen. For instance, words in menus or dialog boxes appear in **bold**. Here is an example: "Click on the **Next: Permissions** option on the other options, and finally, click the **Create User** button."

> **Tips or important notes**
> Appear like this.

Get in touch

Feedback from our readers is always welcome.

General feedback: If you have questions about any aspect of this book, email us at customercare@packtpub.com and mention the book title in the subject of your message.

Errata: Although we have taken every care to ensure the accuracy of our content, mistakes do happen. If you have found a mistake in this book, we would be grateful if you would report this to us. Please visit www.packtpub.com/support/errata and fill in the form.

Piracy: If you come across any illegal copies of our works in any form on the internet, we would be grateful if you would provide us with the location address or website name. Please contact us at copyright@packt.com with a link to the material.

If you are interested in becoming an author: If there is a topic that you have expertise in and you are interested in either writing or contributing to a book, please visit `authors.packtpub.com`.

Share Your Thoughts

Once you've read *Architecting Vue.js 3 Enterprise-Ready Web Applications*, we'd love to hear your thoughts! Scan the QR code below to go straight to the Amazon review page for this book and share your feedback.

`https://packt.link/r/1801073902`

Your review is important to us and the tech community and will help us make sure we're delivering excellent quality content.

Download a free PDF copy of this book

Thanks for purchasing this book!

Do you like to read on the go but are unable to carry your print books everywhere? Is your eBook purchase not compatible with the device of your choice?

Don't worry, now with every Packt book you get a DRM-free PDF version of that book at no cost.

Read anywhere, any place, on any device. Search, copy, and paste code from your favorite technical books directly into your application.

The perks don't stop there, you can get exclusive access to discounts, newsletters, and great free content in your inbox daily

Follow these simple steps to get the benefits:

1. Scan the QR code or visit the link below

https://packt.link/free-ebook/9781801073905

2. Submit your proof of purchase
3. That's it! We'll send your free PDF and other benefits to your email directly

Part 1: Getting Started with Vue.js

This first part gives you a theoretical and historical background for the rest of the book. It covers Vue.js, Options API, the new Vue.js 3, and Composition API. You will also learn how to create a new Vue app using Vue CLI, and then, we will dive deeper into using Vuex, Vue Router, and Axios to build an enterprise-ready app.

This part comprises the following chapters:

- *Chapter 1, Getting Started with Vue.js 3*
- *Chapter 2, Using Libraries for Large-Scale Applications*

1

Getting Started with Vue.js 3

Before we start learning how to develop enterprise-ready applications with Vue.js 3, you need to understand Vue 3 and the different features it is bundled with to help you navigate through building scalable and enterprise-ready applications.

In this chapter, we will cover the essential aspects of Vue 3 that will directly influence how we develop an enterprise application with Vue.js 3. This background information will put you in a better position to grasp the terms and concepts of Vue 3 and help you understand how to build and scale an enterprise-ready application.

We will cover the following key topics in this chapter:

- Overview of Vue.js
- Introducing Vue.js 3
- Building your first Vue.js 3 app

Once you've worked through each of these topics, you will be ready to get started with building your first enterprise-ready Vue.js 3 application.

Technical requirements

To get started, we recommend that you have a basic knowledge of JavaScript with Node.js installed on your computer and must have built projects using Vue.js before.

Overview of Vue.js

Vue.js is an open source progressive JavaScript frontend web framework used to develop interactive frontend web interfaces. It is a very popular and simplified JavaScript framework that focuses on the view layer of web development. It can be easily integrated into big and enterprise web development projects.

Vue.js is a framework that opens the door for developers to create and manage large and scalable projects with ease, as the code structure and development environment are developer-friendly.

In the next section, we will introduce you to the wonders of Vue 3 and the Composition API.

Introducing Vue.js 3

The official Vue.js 3 version was released in September 2020 with highly documented, highly readable, well-structured resources to help you start using Vue 3. Evan You in his article *The process: Making Vue 3* (`https://increment.com/frontend/making-vue-3/`) mentioned that one of the key reasons for the rewrite was to leverage a new language feature, *Proxy*.

Proxy allows the framework to intercept operations on objects. A core feature of Vue is the ability to listen to changes made to the user-defined state and reactively update the DOM. In Vue 3, using the Proxy feature is the key to resolving the reactivity-related issues in Vue 2.

Most importantly, Vue 3 was completely rewritten in TypeScript and has all the advantages of a modern framework that come with using TypeScript.

In this section, we will explore some of the features and improvements that resonate with building an enterprise application and, most importantly, the new Composition API.

We'll cover the following topics:

- Vue 3 performance
- Tree-shaking support
- The Composition API

These topics give you a glimpse at the features of Vue.js 3 and we will start with what we are already familiar with in Vue in this book.

Vue 3 performance

The performance increase in Vue 3 is excellent for enterprise applications because any lag in the core framework can result in a loss of funds given the gigantic nature of an enterprise project.

Vue 3 has sped up performance by 55% compared to previous versions. Also, the updates are up to 133% faster, which is excellent for developing and testing large enterprise projects before deployment. Also, memory usage is reduced by 54%, cutting down computing costs drastically on enterprise projects.

Tree-shaking support

Tree-shaking is the process of eliminating dead, useless, or unused code, which drastically decreases the build size of an application if you compare this to an enterprise application with thousands of files and—sometimes unknowingly—unused files that can lead to a bloated and heavy project.

Vue 3 supports tree-shaking right out of the box, eliminating unused files and code, thereby decreasing the build size and increasing the project's performance.

The Composition API

The Composition API is an entirely new addition and the most significant change to Vue 3. It requires relearning the concepts and total discarding the Options API used in Vue 2. While the Composition API advances, the previous Options API will continue to be supported. In this book, we use the Composition API because of the readability and performance improvements that come with it.

Why the Composition API?

When building a simple application, the component-based architecture alone has proven to be the best approach to developing such an application where individual components can be reused to improve maintainability and flexibility.

However, when building enterprise-ready applications with hundreds of components, from collective experience, it is proven that component-based architecture alone might not be enough, especially when your application is getting big but sharing and reusing code even within components becomes very important, and thus the introduction of the Composition API.

Code example

Let's imagine we are building an enterprise to-do application with unique features such as filters and search capabilities. Using the Options API, we will approach this project using the traditional `data`, `computed`, and `watch` methods.

The following code block shows how to create and manage a Vue component using the Options API from Vue 2:

```
// src/components/TodoRepositories.vue

export default {
  components: { RepositoriesFilters, RepositoriesSortBy,
               RepositoriesList },
  props: {
    todo: {
      type: String,
      required: true,
    },
  },
  data() {
```

```
      return {
        repositories: [], // 1
        filters: {}, // 3
        searchQuery: '', // 2
      }
    },
    computed: {
      filteredRepositories() {}, // 3
      repositoriesMatchingSearchQuery() {}, // 2
    },
    watch: {
      todo: 'getTodoRepositories', // 1
    },
    mounted() {
      this.getTodoRepositories() // 1
    },
    methods: {
      getTodoRepositories() {
        // using `this.Todo` to fetch Todo repositories
      }, // 1
      updateFilters() {}, // 3
    },
}
```

The preceding component handles many responsibilities, as you can see in the following points:

- Getting the Todo repository from an external API and refreshing it on user changes
- Searching the Todo repository using the searchQuery string
- Filtering the Todo repository using the filters object

Organizing your component's logic as in the previous example works perfectly, but at the same time poses a huge challenge to readability and maintainability for larger and enterprise projects with bigger components' logic.

Wouldn't it be perfect if we could collocate code related to the same logical concern? That's exactly what the Composition API enables us to do.

Let's rewrite the same component using the Composition API to see the improvement and readability benefits gained by using it:

```
<script setup>

import { fetchTodoRepositories } from '@/api/repositories'
import { ref, watch, computed } from 'vue'

const props = defineProps({
    todo: {
        type: String
        default:""
    }
})


  const repositories = ref([])
  const getTodoRepositories = async () => {

    repositories.value =
        await fetchTodoRepositories(props.todo)
  }

  getTodoRepositories()

  // set a watcher on the Reactive Reference to user todo
  // prop
  watchEffect(getTodoRepositories)

  const searchQuery = ref('')
  const repositoriesMatchingSearchQuery = computed(() => {
    return repositories.value.filter(
      repository =>
          repository.name.includes(searchQuery.value)
    )
```

```
    })



</script>
```

The Composition API is a great addition, especially for developing enterprise-ready applications. We can move the `computed`, `mounted`, and `watch` lifecycle hooks into a standalone composition function and import it into the script with `setup`, making it readable, flexible, and maintainable. To learn more about the Composition API, visit the official documentation (`https://v3.vuejs.org/guide/composition-api-introduction.html#why-composition-api`), which is outside the scope of this book.

So far, we have covered an overview of Vue 3 and the newly introduced features of Vue that are handy for building enterprise-ready and scalable production-grade applications. We have also covered the basics of the Composition API to foster your understanding of building your modern enterprise application with Vue 3.

In the next section, we will put your knowledge to the test by learning how to build your first Vue 3 application using Vite as the build tool.

According to the official documentation (`https://vitejs.dev/guide/`), Vite is a build tool that aims to provide a faster and leaner development experience for modern web projects. It is based on Rollup, and it's configured to support most sensible defaults for modern JavaScript frameworks.

Building your first Vue.js 3 app

Vue.js can be integrated into projects in multiple ways depending on the requirements because it is incrementally adaptable.

We will create a completely blank new Vue 3 project, or you can use the migration guide (`https://v3.vuejs.org/guide/migration/migration-build.html#overview`) to migrate your Vue 2 project to Vue to follow along.

In this section, we are going to cover how to build our Vue 3 application using the Vite **command-line interface (CLI)**.

Creating a Vue 3 app with Vite

To create our first Vue 3 application, we will use the recommended **Vite** web development tool. Vite is a web development build tool that allows for lightning-fast code serving due to its native ES Module import approach.

In this book, we will be building an enterprise-ready Pinterest clone project, and all the backend data management of the project will be developed and hosted with **Strapi**.

Type along with these simple commands:

```
npm init @vitejs/app pinterest-app-clone
cd pinterest-app-clone
npm install
npm run dev

// If you're having issues with spaces in username, try using:
npx create-vite-app pinterest-app-clone
```

The preceding commands will create a `pinterest-app-clone` folder with Vue 3 installed and set up properly. Once done, open your favorite browser and access the web page with `localhost:3000`. This is what the web page will look like:

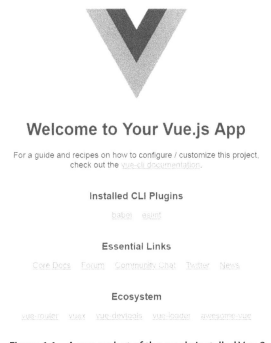

Figure 1.1 – A screenshot of the newly installed Vue 3

In this section, we explored Vue 3, the Composition API, and how to get started building your first application with Vue 3. In the next section, we will learn about the Strapi CMS that we will use for data and content management.

What is the Strapi CMS?

Strapi is an open source headless CMS based on Node.js that is used to develop and manage content or data using RESTful APIs and GraphQL.

With Strapi, we can scaffold our API faster and consume the content via APIs using any HTTP client or GraphQL-enabled frontend.

Scaffolding a Strapi project

Scaffolding a new Strapi project is very simple and works precisely like installing a new frontend framework. Follow these steps to scaffold a new Strapi project:

1. Run either of the following commands and test them out in your default browser:

    ```
    npx create-strapi-app strapi-api --quickstart
    # OR
    yarn create strapi-app strapi-api --quickstart
    ```

    The preceding command will scaffold a new Strapi project in the directory you specified.

2. Next, run `yarn build` to build your app and, lastly, `yarn develop` to run the new project if it doesn't start automatically.

The `yarn develop` command will open a new tab with a page to register your new admin of the system:

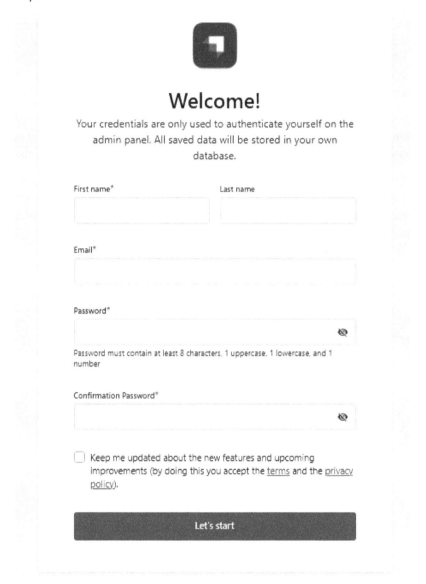

Figure 1.2 – The registration page

3. Go ahead and fill out the form and click on the **Submit** button to create a new admin.

As we progress in this book, we will customize our Strapi backend instance to reflect Pinterest data modeling.

Summary

This chapter started with an overview of Vue.js and why Vue.js can be used to develop enterprise-ready applications. We discussed the latest release of Vue.js and how it improves the performance aspect of the framework by introducing a tree-shaking feature right out of the box. We then introduced the Composition API, a Vue 3 feature that improves the readability, maintainability, and scalability of Vue 3 for building and deploying enterprise applications. We also looked at creating our first Vue 3 application using Vite and the fundamental reasons for using Vite instead of the other available options.

Finally, we introduced the Strapi CMS, the backend stack and a headless CMS for building and modeling backend applications and APIs. With Strapi, we will only focus on building and scaling our enterprise frontend Pinterest-clone application using Vue 3 while the Strapi CMS handles the backend.

In the next chapter, we will dive deeper into using Vuex, Vue Router, and Axios to build an enterprise-ready app. You will learn how to properly utilize these libraries to develop large-scale applications with maintainability and scalability, and by the end of the chapter, you will have learned how to set up your backend with Strapi and connect it to Vue 3.

2

Using Libraries for Large-Scale Applications

Before you start learning how to use different libraries to develop large-scale and enterprise-ready applications with Vue.js 3, you need to understand these individual libraries and the various features they are bundled with to help you navigate through building scalable and enterprise-ready applications.

In this chapter, we will cover the essential aspects of Vuex, Axios, and Vue Router and how to integrate them with Vue 3 to develop an enterprise application. This background information will put you in a better position to grasp the terms and concepts of these libraries and help you understand how to build and scale an enterprise-ready application.

We will cover the following key topics in this chapter:

- Exploring large-scale Vuex
- Structuring with the repository pattern
- Creating a repository folder
- Structuring Vue navigation with Vue Router

Once you've mastered each of these topics, you will be ready to get started with building your first enterprise-ready application with Vue 3.

Technical requirements

To get started with this chapter, we recommend reading through *Chapter 1*, *Getting Started with Vue.js 3*, and its overview of Vue 3 and the Composition API, which will be intensively used in this chapter.

Exploring large-scale Vuex

Vuex is the state management library for Vue applications. It serves as a central store for all the components in a Vue application. It is also a library implementation tailored specifically to Vue.js to take advantage of its granular reactivity system for efficient updates.

Significant benefits can be derived when using Vuex for the state management of a Vue application. Still, it can easily be misused and overwhelmed if not adequately structured—especially when building a large-scale enterprise application—due to the size of the project and the number of components and features that will be introduced in the project.

To tackle this structure problem, we will introduce you to different structures to arrange your Vuex store and the law of predictability discussed in *Chapter 4, Architecture for Large-Scale Web Applications*, to accommodate large-scale Vue applications.

In this section, we will discuss Vuex states, getters, mutations, and actions.

Practically, the usual way to structure your Vuex store is to have every piece of code inside a single `index.js` file called a **single state tree**. This method works perfectly for a small project and helps to avoid navigating through different files to find a single method.

However, when developing an enterprise project using Vuex, using single state trees becomes very bloated and difficult to maintain.

To reduce this large file and split the file into different features, Vuex allows us to divide our store into **Vuex modules**.

Before we dive in, the Vue community has introduced a new state management system called Pinia that fixes the problems of Vuex and is directly compatible with Vue 3. As of the time of writing, Pinia was still in the development and beta phase. You can learn more about Pinia and how to integrate it into your Vue 3 application here: `https://pinia.vuejs.org/`.

Vuex modules

A Vuex module is a way to split our store based on features, where each module can contain its states, getters, actions, mutations, and even nested modules.

This method allows us to split our store into features and create different files and folders to arrange the store correctly.

We'll learn how we can split our store into features in the next subsection.

Assuming our *Pinterest* application will have different states such as photos, users, comments, and so on, we can split the store into separate modules as opposed to having it in a single file.

Using Vuex modules

As stated in the previous section, using Vuex modules comes with great benefits, and we will stick to it throughout this book. Firstly, let's look at the folder structure of our Vuex module store:

```
const moduleA = {
  state: () => ({ ... }),
  mutations: { ... },
  actions: { ... },
  getters: { ... }
}

const moduleB = {
  state: () => ({ ... }),
  mutations: { ... },
  actions: { ... }
}

const store = createStore({
  modules: {
    a: moduleA,
    b: moduleB
  }
})

store.state.a // -> `moduleA`'s state
store.state.b // -> `moduleB`'s state
```

As you can see in the preceding code block, we have created different modules to wrap our Vuex state, actions, and mutations respectively. It can be useful to structure our project into different features.

Now that we understand how to structure our Vuex store for enterprise projects, let's discuss how to access and manage the store from a component in Vue in the next section.

The Vuex state

First and most importantly, let's discuss states and how we can manage the state of a module Vuex store.

The Vuex state is the data you stored inside your Vuex store and can be accessed anywhere in your Vue application.

The Vuex state follows the single-state-tree pattern. This single object contains all your application-level states. It serves as the "single source of truth". But, since we're adopting modularity to manage our enterprise-ready application, we are going only to learn how to access and manage our module states.

The following code snippet shows how to create a simple Vuex state:

```
// initial state
const state = () => ({
  photos: [],
})
```

In addition, you can access a Vuex store outside of components. For example, you access Vuex inside of Vue services, helper files, and so on. However, in the next section, we will explore different ways to access our state in the components.

Accessing state without mapping

Assuming this is our store for all photos in our *Pinterest* application and we have that photos state, how do we access it in our components?

To access the Photos array, we will use our module name with the store name, as shown in the following code snippet:

```
const Photos = {
  template: `<div v-for="(photo, index) in photos"
             :key="index"> <img :src="photo.url"></div>`,
  computed: {
    photos () {
      return this.$store.photos.state.photos
    }
  }
}
```

The previous code snippet shows how to access a moduled store by creating a new Photos component and displaying all the photos in the photo's state.

To access the `Photos` state array, we used the name of the module it belongs to and accessed the state property *before* the `photos` array.

Next, we're going to explore how to access the Vuex state using the mapping approach.

Accessing the state with mapping

The best way to access the store is to use **Vuex state maps**, (https://vuex.vuejs.org/guide/state.html#the-mapstate-helper), which we will use throughout this book. You can go to the official Vuex documentation (https://vuex.vuejs.org/) to learn more.

Using Vuex mappers is great when your components need to make use of multiple store-state properties or getters. Declaring all these states and getters can get repetitive and verbose, and that's exactly what a Vuex mapper tends to solve.

Let's take an example of creating a simple `Photos` component and using the Vuex state to display different images:

```
import { mapState } from 'vuex'
const Photos = {
  template: `<div v-for="(photo, index) in photos"
                :key="index"> <img :src="photo.url"></div>`,
  computed: mapState({
    photos: state => state.photos.photos,
    }
  })
}
```

The preceding snippet creates a `Photos` component, loops through the data from our Vuex state, and displays the images in the store.

There you have it.

We will use this method in further discussion with actions, mutations, and getters. You should *never* forget the names of your modules.

We now have a fair understanding of the Vuex state and modules and how we will structure our enterprise and large-scale Vuex application for easy maintainability and accessibility.

Let's discuss getters and how we can manipulate our Vuex state using **Vuex getters** and map getters in the next section.

Vuex getters

Vuex getters are very useful for manipulating the Vuex state. Sometimes, you might want to filter or sort the Vuex state before returning the data to the components.

Vuex allows us to create getters that manipulate the state just as computed properties in Vue will do. It also caches the result and updates the cache when data changes.

In each module, we will define its specific getters to manipulate the state in that module. For example, we will create a getter to filter photos based on user ID in our `Photos` module, like so:

```
getters: {
  getPhotoByID: (state) => (id) => {
    return state.photos.find(photo => photo.id === id)
  }
}
```

The preceding code snippet shows how to create a getter that filters all the photos added by a particular user whose ID is passed to the `getters` method.

Next, let's access the getter in our component using the `map` helper. See the following code snippet:

```
...mapGetters({
  // map `this.getPhotoByID` to
  // `this.$store.getters.getPhotoByID`
  getPhotoByID: 'getPhotoByID'
})
```

Vuex getters are a great way to manipulate and manage Vuex states before they go out to our components and can come in handy for filtering, sorting, updating, and deleting records from our Vuex state.

Next, we will discuss **Vuex mutations** and how we can use them in an enterprise-ready application.

Vuex mutations

The only way to change the state of a Vuex state is by committing a mutation.

A Vuex mutation is similar to an event. It takes a string called `type` and a function called `handler`. The `handler` function is where you perform the mutation.

Assuming we're still working with our *Pinterest* photo store, we can add a new `Photo` object to our state with the following code snippet:

```
const store = createStore({
  state: {
    photos: []
  },
  mutations: {
    ADD_NEW_PHOTO (state, photo) {
      // mutate state
      state.photos.push(photo)
    }
  }
})
```

Next, we will look at accessing the Vuex mutation module. We can access it using Vuex map helpers, which is the recommended way for enterprise projects:

```
import { mapMutations } from 'vuex'
export default {
  methods: {
    ...mapMutations({
      addPhoto: 'photos/ADD_NEW_PHOTO'
    })
  }
}
```

Lastly, we can call the `addPhoto()` method anywhere in our component, pass the `Photo` object as the only argument, and let Vuex do its thing.

In addition, the most comprehensive place to use Vuex mutations is in **Vuex actions**. In the next section, we will discuss Vuex actions in detail and demonstrate how they can be used in enterprise applications.

Vuex actions

Vuex actions are similar to Vuex mutations, but instead, they are asynchronous and are primarily used to commit Vuex mutations.

Vuex actions can make API calls to our backend server and commit the response to our Vuex state using Vuex mutations.

Traditionally, to make an API call with Vuex actions, we will do it directly inside the store, as the following code snippet shows.

Using our *Pinterest* photo example, we will have a store similar to this one:

```
const store = createStore({
  state: {
    photos: []
  },
  mutations: {
    ADD_NEW_PHOTO (state, photo) {
      state.photos.push(photo)
    }
  },
  actions: {
   async getPhoto (context, id) {
      const photo = await Axios.get('/photos/'+id);
      context.commit('ADD_NEW_PHOTO', photo)
    }
  }
})
```

Next, to dispatch the action in our component, we will stick with using Vuex maps to dispatch the action and retrieve a new photo corresponding to the ID passed into the getPhoto() method:

```
import { mapActions } from 'vuex'
export default {
  methods: {
    ...mapActions({
      getPhoto: 'photos/getPhoto' // map `this.getPhoto()`
      // to `this.$store.dispatch('photos/getPhoto')`
    })
  }
}
```

So far, we have covered a lot on building large-scale applications with Vuex, and we have elucidated Vuex modules, states, getters, mutations, and actions and how to apply them in building enterprise-ready applications.

To further solve the problem of structure, we will introduce you to the use of the **repository pattern** in arranging your Vuex store, structuring all your API calls into a repository, and accessing them in your Vuex actions.

Structuring with the repository pattern

When building a large-scale, enterprise-ready Vue application, you must get the project's structure right from the ground up.

Separating your Vuex store into individual modules based on an application's features is excellent and provides direct access to files, making debugging and maintenance a breeze.

Using this method alone poses a problem. Your Vuex actions become extremely large with many API calls, extracting API data, and handling errors all happening in the Vuex actions.

Introducing the repository pattern helps eliminate this bloated code base and separates the API calls and management from Vuex.

In this section, we will first get an overview of the repository pattern. Then, we will create a repository folder for our Vue application.

Firstly, before we explore how to use the repository pattern in Vuex, let's get a clear overview of the repository pattern and what can be achieved with it.

Overview of the repository pattern

The repository pattern is a significant pattern used in creating an enterprise-level application, either the frontend or the backend of any enterprise application.

It restricts us from working directly with data in the application and creating a new layer for database operations, business logic, and the application UI.

The following is a list of a few reasons you should use the repository pattern in your frontend development, especially when building enterprise applications:

- The data access code is reusable everywhere across the entire project

- It is effortless to implement the domain logic

- You can unit-test your business logic quickly without any form of tight coupling

- It aids in the decoupling of business logic and the application UI

Dependency injection (DI) is good when writing testable enterprise code, and repository patterns help you achieve DI even in your frontend projects.

> **DI**
>
> DI is a programming technique that makes a class independent of its dependencies.

In the repository pattern, you write an encapsulated code base by hiding the details of how your data is retrieved and processed in your Vuex store.

To implement the repository pattern, we will follow the *Consuming APIs Using the Repository Pattern in Vue.js* article I wrote (https://medium.com/backenders-club/consuming-apis-using-the-repository-pattern-in-vue-js-e64671b27b09).

To consume our backend APIs in Vue.js using the repository pattern, let's demonstrate doing so with an example. Let's assume we have a Vuex store action making different API calls, such as the one in the following code snippet, and we want to implement the repository pattern on it:

```
actions: {
  async all({ commit }, { size = 20, page = 0 }) {
    const response = await
        Axios.get(`/photos?size=${size}&page=${page}`);
    const { data } = response;

    if (data.success) {
      commit("STORE_PHOTOS", data.photos);
    } else {
      commit("STORE_ERROR", data.message);
    }
  },

  async getPhoto({ commit }, id) {
    const response = await Axios.get('/photos/'+id);
    const { data } = response;

    if (data.success) {
      commit("STORE_PHOTO", data.photo);
    } else {
      commit("STORE_ERROR", data.message);
    }
```

```
    },
  },
```

Now, we can follow the upcoming steps to improve the Vuex store by implementing the repository pattern.

Creating a repository folder

First, create a folder in the root directory or `src` folder by running the following command:

```
mkdir repositories
```

We will call ours `repositories`. This folder will contain all your repositories and the HTTP client configurations.

Creating a clients folder

We will create a `Clients` folder inside the newly created `repositories` folder. What will be inside this folder are the different HTTP clients used.

Sometimes, due to the nature of the project, some projects might require several HTTP clients to make API calls due to different reasons. One can be a fallback if the default refuses to connect.

Hence, creating a `Clients` folder is crucial to configure all the HTTP clients at once. Run the following command to create one:

```
cd repositories && mkdir Clients
```

Creating an xxxClient.js class

You can create a class file corresponding to the HTTP client you are using. The naming is subjective, and for **Axios**, we will create an `AxiosClient.js` file and put it in all default configurations.

> **Axios**
> Axios is a promise-based HTTP client for Node.js and the browser. It can run in the browser and Node.js with the same code base.

Run the following command to create the folder:

```
touch AxiosClient.js
```

In summary, you might want to use many HTTP clients, so you create different xxxClient.js files for each with their specific configuration.

For Axios, these are my default configurations just for this test:

```
import axios from "axios";
const baseDomain = "https://localhost:1337"; //For Strapi
const baseURL = `${baseDomain}`; // Incase of /api/v1;
// ALL DEFAULT CONFIGURATION HERE

export default axios.create({
  baseURL,
  headers: {
    // "Authorization": "Bearer xxxxx"
  }
});
```

You can add more default configurations for Axios in the preceding file and export the Axios instance.

Creating an individual repository class

Next, we will create an individual repository based on the number of features we have in our enterprise application.

For instance, we are building a *Pinterest* clone application, and we are sure the application will have the **Photos** and **Users** features. So, we can start by creating a repository for the mentioned features by running the following command:

```
cd repositories && touch PhotoRepository.js UserRepository.js
```

These repositories will contain all API calls for the individual features. We will start by creating a **Create, Read, Update,** and **Delete (CRUD)** operation for the respective repositories to give us an overview. In contrast, we will update the repositories as we progress along in the book.

Open the PhotoRepository.js file and add the following scripts:

```
import Axios from './Clients/AxiosClient';
const resource = '/photos;

export default {
    get() {
        return Axios.get(`${resource}`);
```

```
    },
    getPhoto(id) {
        return Axios.get(`${resource}/${id}`);
    },
    create(payload) {
        return Axios.post(`${resource}`, payload);
    },
    update(payload, id) {
        return Axios.put(`${resource}/${id}`, payload);
    },
    delete(id) {
        return Axios.delete(`${resource}/${id}`)
    },
    //b MANY OTHER RELATED ENDPOINTS.
};
```

Next, we are going to open the `UserRespository.js` file and add the following scripts:

```
import Axios from './Clients/AxiosClient';
const resource = '/users;

export default {
    get() {
        return Axios.get(`${resource}`);
    },
    getUser(id) {
        return Axios.get(`${resource}/${id}`);
    },
    create(payload) {
        return Axios.post(`${resource}`, payload);
    },
    update(payload, id) {
        return Axios.put(`${resource}/${id}`, payload);
    },
    delete(id) {
        return Axios.delete(`${resource}/${id}`)
    },
```

```
    //b MANY OTHER RELATED ENDPOINTS.
};
```

We have created two repositories for our *Pinterest* clone application, and any API-related code will go into the individual repository.

Creating a RepositoryFactory.js class

Create a `RepositoryFactory` factory class inside the `repositories` folder by running the following command to export all the different individual repositories you may have created so that they're easy to use anywhere across your application:

```
touch RepositoryFactory.js
```

Once done, paste in the following code:

```
import PhotoRepository from './PhotoRepository';
import UserRepository from './UserRepository';

const repositories = {
    'Photos': PhotoRepository,
    'Users': UserRepository
}
export default {
    get: name => repositories[name]
};
```

Now that we have improved our Vuex store by creating repositories, let's see how to use these in the next section.

Using the repository pattern

Let's see how to utilize the repositories we have created in our Vuex store. Open your Vuex store `photos` file created earlier and replace the `getPhoto` action method with the following code to utilize the repository pattern:

```
import Repository from "@/repositories/RepositoryFactory";
const Photos = Repository.get("Photos");
const Users = Repository.get("Users");
```

```
actions: {
  async getPhoto (context, id) {
    const photo = await Photos.getPhoto(id);
    context.commit('ADD_NEW_PHOTO', photo)
   }

  async getUsers(context) {
    const users = await Users.get();
    context.commit('ADD_USERS', users)
   }
 }
```

Using the repository pattern eliminates the need to handle errors, manipulate the data retrieved from the API in the Vuex store, and only return the actual data needed in Vuex. This approach also utilizes the **Don't Repeat Yourself** (**DRY**) principle of software engineering as the repositories can be used across the project by creating a new one.

Structuring doesn't end when you have your HTTP API calls sorted out with repository patterns. It extends to the way you arrange your navigation. The navigation file should not be bloated with a large code base that's difficult to understand.

In the next section, we will arrange and structure our navigation using Vue Router to ensure maintainability and scalability in our enterprise project.

Structuring Vue navigation with Vue Router

When building an enterprise-ready application, it's evident that the navigation system will be massive since there will be many navigations, routes, and pages.

This section will show you how to structure Vue Router in your enterprise project properly. To achieve this, we will use the *split-by-feature* approach to organizing Vue Router so that it's easy to navigate, as we achieved with Vuex earlier in the chapter.

This approach will create a structure where public and private routes will be separated, and more routes can also be separated individually.

The folder structure

The folder will comprise an index file, a public file, and a private file containing all the routes belonging to each category.

In the root of your `src` folder, create a `router` folder and create the following files inside the folder by typing in the following commands one after the other in your terminal:

```
cd src && mkdir router
touch index.js
mkdir routes && cd routes
touch public.js private.js combine.js
```

The current folder structure is pretty straightforward, and we will customize it more as we progress with the book. Here's what each file will contain:

- `index.js`: This file will contain the `beforeEach` logic and assembling of all the other routes
- `public.js`: This file will contain all the public-facing routes that do not need restrictions, such as the login page, registration page, and so on
- `private.js`: This file will contain all the private routes used for authenticated users and many restrictions and metadata
- `combine.js`: This file will combine private and public files and make it easy to concatenate it with the main router file

Next, let's create an index file to contain the setup of our newly created project.

The index.js file

The index file is the powerhouse. Open the `index.js` file and add the following code to export all the routes created in the public and private files:

```
import { createRouter, createWebHistory } from "vue-router";
import routes from '@/router/routes/combine.js'
const routes = [
    {
        path: '/',
        redirect: '/'
    }
  ].concat(routes)

const router = createRouter({ history: createWebHistory(),
```

```
routes });

// ....
// BeforeEach code here
//.....

export default router
```

In Vue Router, there are two different history modes mostly in use when developing Vue applications with Vue Router:

- Hash mode
- HTML5 mode

Hash mode

This uses a # (hash) symbol before the actual URL to simulate a full URL so that the page won't be reloaded when the URL changes. This is possible because the pages or sections after the hash never get sent to the server. The implication is that it does not impact the SEO of the page but is the default setup for Vue Router.

HTML5 mode

As seen in the preceding example, this is created using the `createWebHistory()` function, and it is the recommended method for enterprise and production-ready applications. It requires a bit of tedious configuration on the server for it to work properly.

The combine.js file

This is a single utility file that combines all the routes in a single file to be exported to the main router file. Open the file and add the following code:

```
import publicRoutes from '@/router/routes/public.js'
import privateRoutes from '@/router/routes/private.js'
export default publicRoutes.concat(privateRoutes)
```

After adding the routes to the utility file, we will import them into the `main.js` file, as shown in the next section.

Adding the router to Vue

Lastly, we will add our router to the Vue instance, as shown in the next snippet. Open the `main.js` file and add the following code:

```
import { createApp } from "vue"
import App from "./App.vue"
import router from "./router/index.js"
import store from "./store"


createApp(App).use(router).use(store).mount("#app")
```

As we progress in this book, we will revisit the `public.js` and `private.js` files to add more routes based on the *Pinterest* clone application we develop.

Summary

This chapter started by exploring the different libraries to develop large-scale and enterprise-ready applications with Vue.js 3. We discussed the individual libraries and their different features in detail to foster our understanding of building scalable and enterprise-ready applications. We also covered the essentials of Vuex, discussing how to structure our large-scale Vuex store by splitting Vuex actions, modules, getters, and states using the split-by-feature approach.

Next, we discussed the essentials of **separation of concerns** (**SoC**) by using the repository pattern to split large Vuex actions into individual repositories and make our enterprise application maintainable. The repository pattern is essential in creating an enterprise application, and we demonstrated how to implement it in Vue 3.

Lastly, we discussed how to structure Vue Router to avoid bloated and large router files as it will be difficult to maintain when the project becomes larger. We discussed strategic patterns to split the Vue Router files into different files to enable maintainability and ease of debugging.

In the next chapter, we will dive deeper into scaling the performance of enterprise Vue 3 applications. We will explore different performance and scalability hacks to build an enterprise-ready Vue 3 application, such as asynchronous components' loading/lazy loading, tree shaking, image compression, and so on. You will learn how to properly increase the performance of your Vue 3 application by applying the tricks in the next chapter to develop large-scale applications with maintainability and scalability.

Part 2: Large-Scale Apps and Scaling Performance in Vue.js 3

In this part, you will learn best practices in building large-scale applications, with scalability and performance as first-class citizens. In addition, you will learn and explore different techniques in scaling large application performance.

This part comprises the following chapters:

Scaling Performance in Vue.js 3

This chapter depends solely on the knowledge of the previous chapters, where we explored the different libraries to develop large-scale and enterprise-ready applications with Vue.js 3. This chapter will dive deeper into scaling an extensive Vue application. You will learn how to scale performance with asynchronous lazy loading, image compression, code splitting, tree shaking, and many other tricks to better increase the performance of your Vue.js 3 enterprise-ready application.

We will cover the following key topics in this chapter:

- Why do we need Vue.js performance optimization?

- The primary reasons for poor Vue performance

- Checking your Vue.js application's bundle size

- Optimizing the performance of an enterprise Vue application

Once you've mastered each of these topics, you will be ready to get stuck into building your first enterprise-ready application with Vue 3.

Technical requirements

To get started with this chapter, I recommend you read through *Chapter 1, Getting Started with Vue.js 3*, where you will get an overview of Vue 3 and the Composition API, intensively used in this chapter.

Why do we need Vue.js performance optimization?

In this section, we will learn why performance stability in an application is important and how to develop an application with performance in mind.

Developing an application without taking actionable steps to ensure the stable performance of the application can cost the application a lot. Developing an application that takes a while to load, navigate, submit, or take any user actions will result in losing users, thereby gradually losing on the initial plan of the application.

Suppose the end users are not satisfied with the enterprise application's user experience and load time, Vue.js performance, and efficiency. In this case, the time invested and the lines of code written don't matter; the user might not return to the application.

Here are some different facts from Kinsta that show how poor performance can affect the performance of an enterprise application on the market: `https://kinsta.com/blog/laravel-caching/`.

An online study (`https://kinsta.com/learn/page-speed/#slow-how-slow`) found that it cost Amazon $1.6 billion in sales per year for every 1 second of load lag time.

Another Google study (`https://www.thinkwithgoogle.com/future-of-marketing/digital-transformation/the-google-gospel-of-speed-urs-hoelzle/`) reported that if search results are slow even by a fraction of a second, people will search less. What this means is that a 400-millisecond delay leads to a 0.44% drop in search volume.

A further study shows that four out of five internet users will click away if a video stalls while loading.

The preceding study shows that a slight sluggishness in your web page load time can have a massive impact on your users' experience and the loss of a huge amount of funds.

Now that we know why we need performance stability in our application, in the next section, let's look at the primary reasons behind poor Vue performance.

The primary reasons for poor Vue performance

There are many known reasons behind poor Vue performance, and we will explore the primary and most notable reasons in this section.

The apparent reason for a Vue application slowing down is in the structure. As an enterprise application, it's evident that the bigger the application, the slower the application becomes.

For example, one of the significant reasons behind poor performance in enterprise Vue applications may vary in different projects and how they deal with **server-side rendering** (**SSR**).

The primary reason for poor performance in any Vue **single-page application** (**SPA**) or SSR enterprise application is the bundle size. The larger the bundle size, the slower the Vue performance.

There are other common reasons behind poor performance in enterprise Vue applications, such as the following:

- Not structuring the CSS and JS files properly
- Not using third-party libraries wisely
- Unwanted hits to API requests
- Overlooking code splitting and lazy loading

There are more reasons for poor performance, and we have just listed a few. Before we discuss how to resolve them, let's explore how to check for the bundle size of an enterprise Vue application in the next section.

Checking your Vue application's bundle size

The bundle size is the total size of your Vue application that will be loaded by the browser. The larger the size, the slower your application loads.

There are two different ways to check your Vue bundle size when working with the Vue framework.

Let's go through each of these methods in more detail.

Generating a report

You can use the `build` command with the `--report` flag to generate your application report. This method gives a visual representation of all the packages used and each bundle size. Further, with the information generated from this visual report, you can figure out how to replace any package that takes up more space and size than expected.

Also, note that the `build` command will only build a report when `webpack-bundle-analyzer` is installed.

To generate a report for your application, follow these steps:

1. First, install the package with the following command:

    ```
    npm install webpack-bundle-analyzer
    ```

2. Next, create a script for the command in your `package.json` file:

    ```
    "build-report": "vue-cli-service build --report"
    ```

3. And lastly, execute the following command to generate the report:

    ```
    npm run build-report
    ```

After running the preceding command, a file named `report.html` is created inside the `dist` folder. When you open the file, you will see the following:

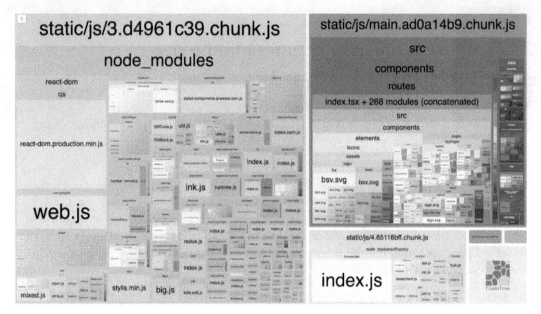

Figure 3.1 – Application bundle-size report

Running the npm build command

Running the `build` method of your Vue application will generate a list of different chunks and bundle sizes. From this information, you can see additional warnings concerning which chunk has a bigger bundle size and how you can improve it. Here's how it looks:

```
warning

entrypoint size limit: The following entrypoint(s) combined asset size exceeds the recommend
Entrypoints:
    app (2.4 MiB)
        css/chunk-vendors.a9a5dc81.css
        js/chunk-vendors.32299ff1.js
        css/app.52da51ea.css
        js/app.bf2b207c.js

File                                        Size            Gzipped
dist\js\chunk-vendors.32299ff1.js           2109.54 KiB     550.18 KiB
dist\js\chunk-fcff55b8.c547ab17.js          658.60 KiB      167.75 KiB
dist\js\chunk-1717e989.941e5795.js          388.94 KiB      114.23 KiB
dist\js\chunk-0fed8959.40c86c37.js          300.52 KiB      100.15 KiB
dist\js\chunk-4834c7cc.5362022b.js          139.10 KiB      41.96 KiB
dist\js\chunk-bad11e74.c18f8684.js          84.90 KiB       26.10 KiB
dist\js\chunk-260bc333.30ddd5c9.js          70.88 KiB       25.09 KiB
dist\js\app.bf2b207c.js                     49.71 KiB       14.96 KiB
```

Figure 3.2 – Chunks and bundle sizes

In this section, we learned why we need Vue.js performance optimization, the primary reasons for poor performance, and the different ways to check Vue.js bundle size.

In the next section, we will learn how to optimize the performance of a Vue application using different standard methods.

Optimizing the performance of an enterprise Vue application

One of the downsides of creating an enterprise application is the size of the application regarding the code base, the data size, and the speed it takes to respond to users' actions.

One solution could be to implement a proper caching mechanism on both the backend and frontend of the enterprise application.

You will agree that it's challenging to develop an application. Still, it is more challenging to create an application with optimized performance or even solve the performance bottleneck of an enterprise application.

In this section, we will look at some tips that you can implement to improve the performance of your enterprise Vue application.

Asynchronous/lazy component loading

We will start with asynchronous/lazy components loading to lessen your challenges to explore Vue. js performance optimization.

Asynchronous/lazy components loading in Vue.js is a term used to describe loading modules/ components when the user needs a module/component. In an enterprise application, it is unnecessary to load all the modules from the JavaScript bundle whenever the user visits the website, as doing so will cause a performance bottleneck.

In enterprise projects, you will agree that there are complex components with many modals, tooltips, and other interconnected components that will slow down the performance of your application if not lazy loaded.

Before we explore how to lazy load components, you can check the actual JavaScript code used on your web page by following these simple steps:

1. Click on **DevTools**. The following screen will appear:

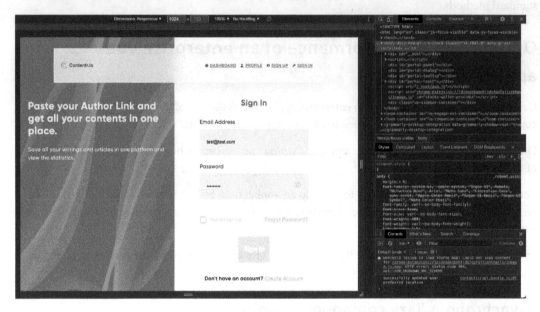

Figure 3.3 – Sample of live DevTools in Chrome

2. Press *Cmd + Shift + P*.

3. Type `Coverage`. Once you type it, the following message will appear at the bottom of the screen:

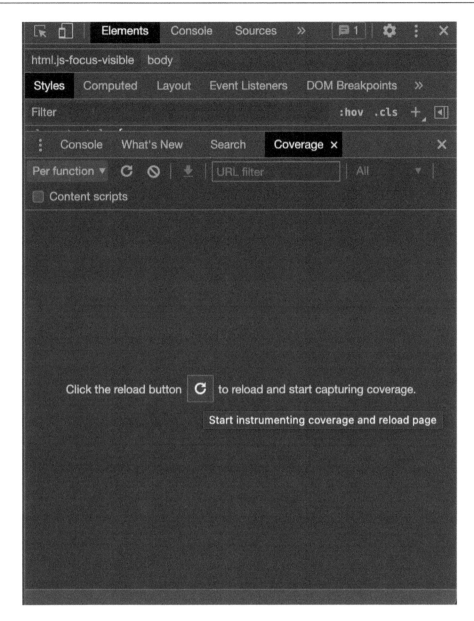

Figure 3.4 – A screenshot showing the Coverage tab

4. Click **Record**.

After recording and analyzing the web page, it will highlight some URLs in red, which shows that the URLs are not in use and can be lazy-loaded:

Figure 3.5 – A screenshot showing different URLs when analyzing with DevTools

If lazy loading is appropriately implemented, the bundle size of your enterprise application can be reduced to 60%, thereby increasing the speed of your application.

Lastly, let's explore how to enforce lazy loading. We can use Webpack dynamic imports over regular imports to separate the chunk of lazily loaded modules.

Traditionally, this is how components/modules are imported in JavaScript:

```javascript
// photo.js
const Photo = {
  testPhoto: function () {
    console.log("This is just a Photo Component!")
  }
}
export default Photo
// app.js
import Photo from './photo.js'
Photo.testPhoto()
```

By importing this module this way, Webpack will create a file named `photo.js` as a node to the `app.js` file in its dependency graph and bundle it together even when the user doesn't need to use the module.

But to improve things a little, we can use dynamic imports or lazy loading to achieve and overcome the performance bottleneck with the previous method. The following code block shows dynamic/lazy loading in action:

```
// app.js
const getPhoto = () => import('./photo.js')

// later when some user action tasks place as hitting the
// route
getPhoto()
  .then({ testPhoto } => testPhoto())
```

Lazy loading in Vue.js is one of the best practices to reduce the bundle size and optimize performance. Vue recommends that bundles should not exceed a size of 244 KiB, but you can also try to optimize your web page to make sure that it's not very slow in performance, even with a bundle size a little higher than recommended. Develop a habit of knowing which modules you don't need unless there's an explicit user action, and download them lazily for better performance.

WebP images and image compression

One of the primary reasons for large bundle sizes and slow applications is that images contribute a lot to an application's bundle size. If images are not correctly optimized, it can increase the loading time of an application when the application renders images of considerable sizes.

There are different ways to achieve image optimization, and we are going to discuss two of the popular methods:

- Compressing images
- Optimizing **content delivery network (CDN)** images

Compressing images

If your enterprise application contains small images in number, these images will be served locally while applying a different compression algorithm to reduce the sizes of each image.

There are thousands of online tools to compress images, and the following is the list of some popular ones:

- TinyPNG
- Compressnow

- Image Compressor
- Adobe Photoshop

Also, the best compression algorithm used to reduce the sizes of an image is the WebP image format (https://developers.google.com/speed/webp), which is developed and maintained by Google.

Optimizing CDN images

A CDN is used to optimize images. It provides transformation features for reducing image sizes by up to 70% without affecting the UI and pixelating. It's also advisable to use a CDN when your enterprise application deals with extensive media use.

The most popular CDN tools for image optimization are **Cloudinary** and **ImageKit**.

Media takes a considerable amount of space in any enterprise application and therefore can cause lagging and slow performance if not optimized and served appropriately.

Code splitting

MDN explains the following (https://developer.mozilla.org/en-US/docs/Glossary/Code_splitting):

> *"Code splitting is the splitting of code into various bundles or components which can then be loaded on demand or in parallel.*

> *As an application grows in complexity or is maintained, CSS and JavaScripts files or bundles grow in byte size, especially as the number and size of included third-party libraries increases."*

When creating an enterprise application, there will always be many routes, files, and bundles that will increase the byte size of the enterprise application. Code splitting is the answer to separating and only loading smaller and on-demand files, thereby increasing the load time of your enterprise application.

Let's our enterprise application has two pages and we implement it with the popular vue-router library, as we have here:

```
```
// routing.js
import Dashboard from './Dashboard.vue'
import Photo from './Photo.vue'

const routes = [
```
```

```
    { path: '/', component: Dashboard }
    { path: '/photo, component: Photo }
]
```

Due to the coding standard in Vue.js, all the components in our script will be downloaded when the user visits any page. This activity causes slow performance due to the number of pages, the complexity of each page, and the large bundle size.

To avoid this issue, we can implement a proper route code splitting that will separate our large bundle into different route bundles, meaning each page will have its small bundle to download when a user visits that page.

With the technique of dynamic imports, rather than importing the components directly as demonstrated previously, we can pass the dynamic route and lazy-load the component, as shown in the following code block:

```
// routing.js
const routes = [
    { path: '/', component: () => import('./Dashboard.vue') }
    { path: '/photo, component: () => import('./Photo.vue') }
]
```

By following this method, you can halve your bundle size. Also, it is important to be sure which components can be used with dynamic imports.

Summary

In this chapter, we dove deeper into scaling an extensive Vue application. We discussed how to scale performance with asynchronous lazy loading, image compression, code splitting, tree shaking, and many other tricks to better increase the performance of your Vue.js 3 enterprise-ready application.

We also covered in detail why performance optimization is needed and what your enterprise application can lose if performance is not deliberately built into the application. We also discussed the reasons for poor performance in an enterprise application and how to fix them.

Next, we discussed how to check our Vue.js application's bundle size, demonstrating this with simple instructions on how to generate package reports using Webpack and commands. We also discussed how to understand the report and discover how to improve an application from the generated report to further boost our enterprise application's performance.

In the next chapter, we will learn how to handle a sizable enterprise-ready project, from managing larger file structures to using a micro frontend architecture. You will also learn how to handle the internationalization and localization of your Vue.js 3 project.

4
Architecture for Large-Scale Web Apps

In the previous chapter, we explored building and scaling large-scale applications in Vue 3. We discussed why we need Vue.js performance optimization, the primary reasons behind poor Vue performance, how to check your Vue.js application's bundle size, and optimizing the performance of an enterprise Vue application using different methods such as asynchronous/lazy component loading, WebP images, and image compression and code splitting.

In this chapter, we will learn how to handle a sizable enterprise-ready project, from managing larger file structures to using the micro frontend architecture. We will also learn how to handle the internationalization and localization of our Vue.js 3 project.

We will cover the following key topics in this chapter:

- File architecture and structure
- Micro frontend architecture
- Internationalization and localization

By the end of this chapter, you will know how to architect large-scale web apps with Vue 3, how to implement structures and file architecture using the law of predictability, and how to use community-recommended packages to inform predictability in your Vue.js 3 enterprise-ready application.

You will also learn how to use micro frontend architecture to your advantage and how to implement an Atomic Design with Storybook to streamline your component directory and make your enterprise project less difficult to understand.

And lastly, you will learn how to add and properly integrate internationalization in to your Vue application and about the benefits this brings.

Technical requirements

To get started with this chapter, I recommend you read through *Chapter 3*, *Scaling Performance in Vue.js 3*, where we elucidate the building and scaling of large-scale applications in Vue 3.

All the code files for this chapter can be found at `https://github.com/PacktPublishing/Architecting-Vue.js-3-Enterprise-Ready-Web-Applications/tree/chapter-4`.

Understanding file architecture and structure

Structuring your project depends solely on the preference of your organization and how easy it is to access files and folders when fixing bugs and adding new features.

In this section, we will explore different principles that will give you an idea of how you can structure your project to incorporate best practices, standards, and easy-to-access files.

What is the most effective way to structure your project to scale and keep it maintainable and extendable the more it grows?

This is a common question in the software development industry but there is no one-size-fits-all method. It all depends on the principle of predictability, as discussed in this article: `https://vueschool.io/articles/vuejs-tutorials/how-to-structure-a-large-scale-vue-js-application/`.

The principle of predictability is simply the ability to go from point A to point B in any code base to intuitively go from a feature request or bug report to the location in the code base where the said task can be addressed. Furthermore, it's the ability to quickly or easily understand a particular code base based on the standard use of community or popular libraries and tools.

To elaborate, when a code base uses standard, community-agreed, and popular libraries or tools, it gives a great developer experience as developers are already familiar with these tools.

In the next section, we will delve deeper into discussing predictability and how we can achieve it in Vue 3.

Predictability in Vue 3

How to achieve predictability in Vue 3 is very simple, as stated previously; it boils down to using Vue 3 standards and style guides.

For example, just imagine buying a new iPhone 13 ProMax in a different size; it will be awkward since you must have certainly predicted the size to stay the same from your reviews.

The approach applies to the mindsets of developers toward a new code base; we expect most libraries, component names, files, and folder structures to follow Vue 3 community standards and style guide with a little adjustment to suit the organization's use case (if any).

So how can we achieve predictability in Vue 3? In the following subsections, we will look at a few ways to achieve standards in your enterprise Vue 3 application.

Community-wide standards for predictability

If you're coming from Vue 2, you should already be familiar with the standards that exist within it. We will discuss adding more Vue 3 specific standards from there.

Vue has the following pages where you can look out for community standards:

- Start by reviewing the official Vue.js style guide (`https://v3.vuejs.org/style-guide/#rule-categories`)

- Always use the scaffolding generated by the Vue **command-line interface** (**CLI**) or Vite in Vue 3 (`https://vuejs.org/guide/quick-start.html`)

- The official Vue.js libraries are found under the *Community Guide* (`https://vuejs.org/about/community-guide.html`)

- Use one of the most popular component frameworks such as Vuetify (`https://vuetifyjs.com/en/`) or Quasar (`https://quasar.dev/`)

Official libraries and component libraries

Using official libraries and component libraries not only brings functionality to your project but also enforces standards and allows you to build applications following standard and generally acceptable patterns according to the Vue community.

For example, Vuex is a state management system that prides itself on implementing a pattern and a library together because it enforces a standard to follow when building Vue applications.

Another great example is Vue Router, which enables developers to build routing systems in ways that are adaptable to other projects.

And the good thing about all of this is when a developer who has built with these libraries is added to a new code base, using these tools, it becomes predictable.

Standard file structure

Another important aspect of project standards is the file structure. The file structure is an arguable aspect of project standards because different organizations and projects use different structures and Vue does not provide detailed documentation specifying a structure.

However, when you use the official Vue CLI, it provides a starting point for creating a standard folder and file structure that is widely used in the Vue.js world, and it's most familiar for Vue developers around the world.

The following code block shows how to create a new Vue 3 project using the official Vue 3 standalone CLI called Vite:

```
npm create vite@latest
npm create vite@latest my-vue-app --template vue
```

The following screenshot shows the official project scaffolding using the Vue 3 official CLI called Vite:

Figure 4.1 – Official Vue CLI file structure

The initial structure used in the preceding screenshot should already be familiar to many developers, therefore making it predictable. Always stick with Vue's initial structure, build on it, and only change it for a good reason.

Recommended component rules

The Vue component directory is where the confusion begins because thousands of files and Vue components can be created, and it becomes very difficult to manage with time.

Adapting your code base to follow the official Vue 3 style guide is a starting point for a predictable code base and you can learn a lot about making your folder and file structure more predictable for developers from there. The style guide provides lots of community-wide standards and best practices for the Vue ecosystem.

Some of the most important points are listed here:

- First, we have the **Single-File Component** (**SFC**) style guide, which states a lot of points to follow, with the important one being that your components should be named in PascalCase.

- Secondly, an SFC (`https://vuejs.org/guide/scaling-up/sfc.html`) should always order the `<script>`, `<template>`, and `<style>` tags consistently, with `<style>` at the end. This is because the `script` and `template` tags are always necessary while the style tag is optional.

- It also states that, when possible, each component should be defined in its own dedicated file (SFC). This is where Storybook or Atomic Design, in general, comes in to play as we will see in the following sections.

- Additionally, component names should always be multi-worded to not conflict with any existing or future HTML elements. Don't create a `Table` component or a `Button` component since there are HTML tags with those names already; you can create a multi-word such as the following `DataTable` or `CustomButton`.

- Most importantly, tightly coupled child components should be prefixed with their parent component's name, such as `TodoListItem` in a `TodoList` component. The method also helps in debugging, as developers can easily spot components with names in error messages.

Vue.js has a full style guide at `https://vuejs.org/style-guide/` with a number of other standards that will help your project be more predictable for a community-wide audience of developers.

Recommended community-wide standards for predictability

Over the years, the Vue community has developed and argued on numerous different standards that should be used by Vue developers for a more predictive code base.

In the following subsections, we will discuss a handful of these standards and how to implement them in your enterprise project.

A flat component directory

A flat component directory entails giving a specific naming convention to your Vue components and your team, sticking with that convention throughout application development.

You can use a single or nested directory structure, but the naming convention should stay the same. The next two screenshots show different ways to implement the flat component directory. The following screenshot shows a single flat component directory:

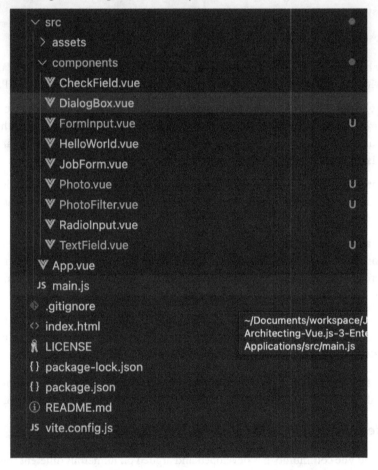

Figure 4.2 – Single flat component directory

The following screenshot shows a nested flat component directory:

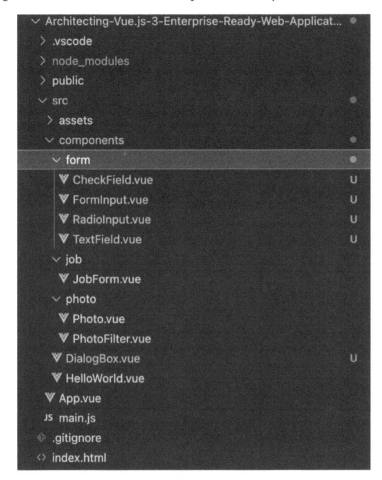

Figure 4.3 – Nested flat component directory

Standardized route/page naming convention

Another important improvement to the principle of predictability is having proper and well-defined route/page naming conventions throughout your team and project.

For instance, using routes/page naming conventions used in Laravel or AdonisJS makes it easy for developers who have prior knowledge of these frameworks to quickly predict the code base. The same is applicable if you define your custom convention and stick with it through your team. It allows new members to easily predict and understand your code base.

The following screenshot shows how you can structure your routes to be predictable based on routing standards from Laravel and AdonisJS:

Path	Route and Component Name	What it Does
/users	UsersIndex	List all the users
/users/create	UsersCreate	Form to create the user
/users/{id}	UsersShow	Display the users details
/users/{id}/edit	UsersEdit	Form to edit the user

Figure 4.4 – Showing the pattern that can be adopted

You should always reference your routes properly with their name when using them in router links and programmatically for more consistency and flexibility.

For example, see the following:

```
<router-link :to="{name: PhotosIndex}">Photos</router-link>
```

Also, note that not all routes will fit this pattern exactly, as some routes will be cruddier than others. If this happens, a good recommendation is to continue using PascalCase for your route names for consistency.

A more comprehensive file structure

Using the basic file structure from the Vue CLI is a great starting point for predictability and can be extended from there to include other files and directories in a way that standardizes our enterprise project for better predictability.

The following screenshot shows how to extend the file structure to include other necessary files and directories:

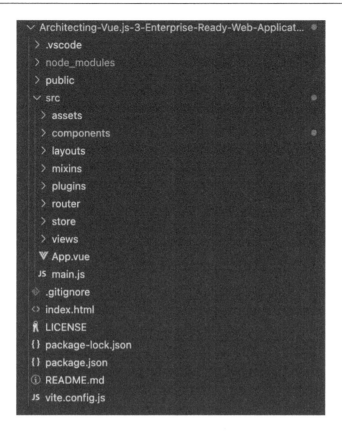

Figure 4.5 – Extending the file structure from the default Vue CLI structure

The additional files and folders will solely depend on your team, organization, or project, but the additional folders in *Figure 4.5* are the structure we have defined for the project we are building, and extending it from the default Vue CLI structure makes it more predictable.

Also, providing a README.md file (https://changelog.md/) in the root of a standard directory explaining the purpose of the directory and any rules for what should be included in it or how to use the directory files is very useful. This comes in handy, especially for those standards that aren't community-wide standards.

While we tend to make our code base predictable enough for developers, no matter how well the project uses community-wide standards and Vue style guides, there are cases where we need to define specific files and folders that are generic to our project or team.

While creating a predictable code base is great for larger projects and teams using the steps and patterns discussed in the previous sections, there is still a lot to explore, and in the next section, we will explore different patterns, architectures, and structures that can be used to structure your larger-scale enterprise projects.

Different frontend architectural patterns

In this section, we will explore different architectural patterns we can use to structure our enterprise Vue 3 applications.

Micro frontend architecture

Micro frontend is the first architecture that comes to mind when it comes to structuring enterprise frontend projects. As expressed in the official documentation, it extends the concept of microservices in the backend to the frontend world.

The concept of a micro frontend comes from the buzzword microservices (`https://martinfowler.com/articles/microservices.html`) used in a backend web application to split gigantic blocks into a smaller, more manageable code base.

This approach to software development makes it easier for teams to manage, maintain, and deploy larger and enterprise applications faster.

This concept, which has changed the way backend applications have been developed over many years, is introduced into frontend projects in the form of micro frontends.

According to Martin Fowler (`https://martinfowler.com/articles/micro-frontends.html`), *"Micro Frontend is an architectural style where independently deliverable frontend applications are composed into a greater whole."*

In recent years, since the initial adoption, there has been tremendous adoption of this concept in larger projects, thereby bringing the benefits of microservices into frontend projects.

The following are some key benefits that come with implementing the micro frontend architecture:

- It comes with more scalable organizations with decoupled and autonomous teams
- It brings smaller, more cohesive, and maintainable code bases
- The provides the ability to upgrade, update, or even rewrite parts of the frontend in a more incremental fashion

As much as there are tremendous benefits to using this architecture in your enterprise project as outlined in the official documentation, the pattern requires a steep learning curve, a higher number of team expatriates, and a large number of team members.

The following diagram shows an end-to-end example of teams using a micro frontend for the Pinterest demo application:

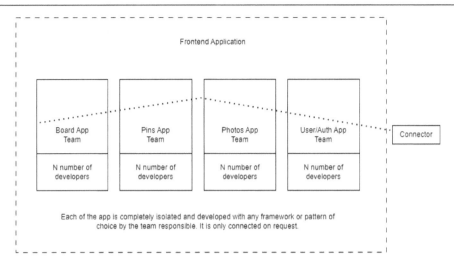

Figure 4.6 – An end-to-end example of teams using a micro frontend

Here's a diagram of how structuring the Pinterest demo application will look using the micro frontend architecture:

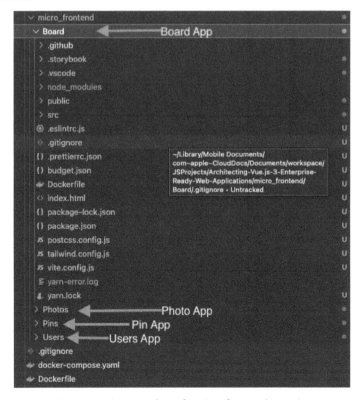

Figure 4.7 – A screenshot of a micro frontend in action

From the screenshot, we can easily separate each of the features into a different service and have a dedicated team of frontend engineers to work on it.

Micro frontend is one of the best architectural patterns to use in enterprise applications because the core ideas behind a micro frontend are having isolated team code, it being technology-agnostic, ownership, and so on. These features make developing enterprise applications a breeze. However, other patterns are also widely used, and we will explore them in the next section.

Atomic Design

Atomic Design (`https://bradfrost.com/blog/post/atomic-web-design/`) is a methodology for crafting design systems. Brad Frost first introduced it for creating scalable design systems using ideas from chemistry.

From chemistry class, we know that matter comprises atoms that bond together to form molecules, which in turn combine to form more complex organisms and ultimately create all the matter in the universe.

Similarly, we can break down our components into fundamental building blocks and work up from there. These building blocks can be divided into five components from the chemistry example, as listed here:

- Atoms
- Molecules
- Organisms
- Templates
- Pages

This diagram from Rohan Kamath (`https://blog.kamathrohan.com/atomic-design-methodology-for-building-design-systems-f912cf714f53`) gives a clear illustration of the Atomic Design elements:

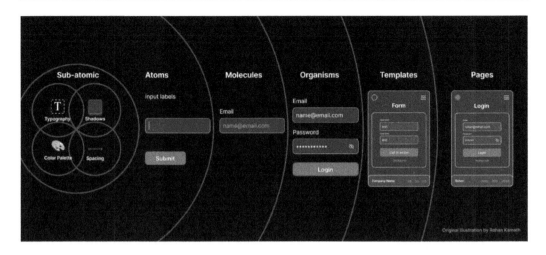

Figure 4.8 – Atomic Design elements explained (source: https://blog.kamathrohan.
com/atomic-design-methodology-for-building-design-systems-f912cf714f53)

Let's explore each of these components to understand them.

Atoms

In science class, we learned that atoms are the basic building blocks of matter. But when applied to web interfaces, atoms are the HTML tags such as `input`, `label`, and so on. They can also be customized to include abstract elements such as color palettes, fonts, or animations.

Atoms are not very useful on their own except when combined with other elements to form molecules.

Molecules

When we start combining atoms, things start to get a little interesting and tangible.

The smallest unit of a compound is called a molecule, and it comprises groups of atoms bonded together. In web interfaces, these molecules take on their own properties and serve as the backbone of any design system.

For instance, a form input, label, or button is not very useful as standalone functionality, but when combined as a form, they become very useful because they can actually do something. Furthermore, by combining atoms to form useful components, these components become reusable and can also be combined to form organisms.

Organisms

An organism is the combination of different molecules used to form a relatively complex and distinct section of a component.

An organism is designed to consist of different or similar molecule types. For instance, a molecule can consist of a primary navigation, a list of social media channels, a search form, and a logo.

The wonderful part of building an organism from molecules is that it encourages creating standalone, portable, or reusable components.

Templates

Templates will already be familiar to the web development world; they are predefined groups of organisms stitched together to form a page. In templates, the designs start to come together, and the layout of the page becomes structured and visible.

Each template contains all related abstract molecules, organisms, and atoms in some cases. Since templates are visible pages or part of a page, clients can start to see the final design.

With templates, you can create different versions of your design, whether high fidelity, low fidelity, and so on. Templates are more HTML wireframes and can also become the final deliverable.

Pages

A page is a specific instance of a template; in some cases, a complex page can contain more than one template combined to form a bigger page.

A page gives an accurate depiction of what the user will ultimately see, and they are the highest level of fidelity and most tangible. It is typically where most time is spent, and more reviews revolve around it.

In Vue.js, pages represent the different routes users access when navigating your application.

Using Atomic Design principles gives us the ability to traverse from abstract pages or templates to concrete ones. Because of this, we can create systems that promote consistency and scalability while simultaneously showing things in their final context.

In this book, we will learn how to use the Atomic Design pattern to structure our enterprise project and use Storybook for the design system.

> **Tip**
> A design system is a set of interconnected patterns and standards to manage design at scale by reducing redundancy while creating a shared language and visual consistency across different pages and channels.

Storybook

Storybook can be implemented with any architectural pattern, such as Atomic Design, to build component-driven **user interfaces** (**UIs**) faster. According to the official website (https://storybook.js.org/), Storybook is an open-source tool for building UI components and pages in isolation. It streamlines UI development, testing, and documentation.

Storybook allows us, the developers, to create and test components in isolation.

In the next section, we will learn how to implement Storybook into our project and start using the Atomic Design principles to create a maintainable Vue.js 3 project.

Implementing Storybook in Vue.js 3

Visit the official documentation for Vue implementation (`https://storybook.js.org/docs/vue/get-started/introduction`) to follow along with the implementation. In Storybook, everything revolves around stories. A story describes the state of a rendered component and captures everything a component should/can do when rendered.

Installing Storybook

You can use the Storybook CLI to install it in a single command by running it inside your existing Vue.js project's root directory.

Storybook will look into your project's dependencies during installation and provide you with the best configuration available.

Next, depending on your framework, first, build your app and then check that everything works by running the following command:

```
npx sb init
npm run storybook
```

The preceding command will start a new development server and open a browser window showing you a welcome screen:

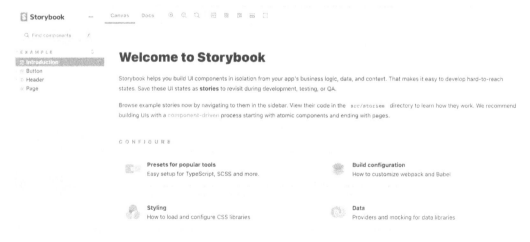

Figure 4.9 – The welcome screen

Creating a story is as easy as telling the computer what a particular component will do, the properties needed to carry out the task, and the different designs a particular component can have. In the next section, we will explore how to create a story in Storybook.

Creating a story

Before we delve into creating stories (components), let's make sure we're in sync with our folder structure for this project using Atomic Design and Storybook.

The following screenshot shows the complete folder structure for implementing Atomic Design and the stories folder for Storybook:

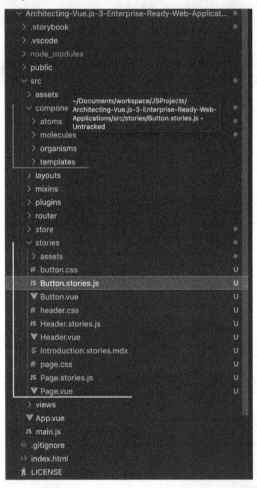

Figure 4.10 – A screenshot showing the Vue 3 component and the Storybook structure

From the screenshot example, we have restructured our Vue.js project to use the Atomic Design principle and folder structures (the red indicator), and Storybook added the `stories` folder (the yellow indicator) to help us understand how to write our own stories. We can delete the `stories` folder and follow the pattern in *Figure 4.10* to create our story inside the `component` folder.

Now, we can start creating stories; remember, a story has to depict a particular action or group of related actions.

Here is a story we created for the `Button` component we will use throughout the project:

```
import MyButton from "./Button.vue";


// More on default export:  https://storybook.js.org/docs/vue/
writing-stories/introduction#default-export
export default {
  title: "/Button",
  component: MyButton,
  // More on argTypes:
  // https://storybook.js.org/docs/vue/api/argtypes
  argTypes: {
    backgroundColor: { control: "color" },
    onClick: {},
    size: {
      control: { type: "select" },
      options: ["small", "medium", "large"],
    },
  },
};


// More on component templates: https://storybook.js.org/docs/
vue/writing-stories/introduction#using-args
const Template = (args) => ({
  // Components used in your story `template` are defined
   // in the `components` object
  components: { MyButton },
  // The story's `args` need to be mapped into the template
  // through the `setup()` method
  setup() {
    return { args };
```

```
  },
  // And then the `args` are bound to your component with
  // `v-bind="args"`
  template: '<my-button v-bind="args" />',
});

export const Primary = Template.bind({});
// More on args: https://storybook.js.org/docs/vue/writing-
// stories/args
Primary.args = {
  primary: true,
  label: "Button",
};

export const Secondary = Template.bind({});
Secondary.args = {
  label: "Button",
};

export const Large = Template.bind({});
Large.args = {
  size: "large",
  label: "Button",
};

export const Small = Template.bind({});
Small.args = {
  size: "small",
  label: "Button",
};
```

For instance, a button will be created in the atoms directory because it is a single element, though it can have different properties and actions, such as being a blue button, white button, clickable button, disabled button, and so on. It's still a button.

From the story, we can see that the Button component will have two sizes (small and large), also, it will have two designs, which are primary and secondary, and additionally, it will accept two properties, namely primary and label.

You can learn how to write a story to test the properties and actions of the button using the official documentation (`https://storybook.js.org/docs/vue/get-started/whats-a-story`).

Once you have created all your components and stories, you should have a directory like the one shown in the following screenshot:

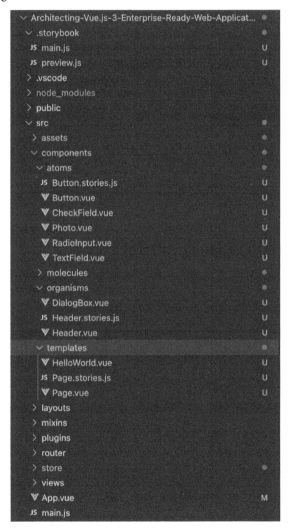

Figure 4.11 – A screenshot showing a complete directory including Storybook stories

Furthermore, with your project arranged like this, it becomes easy for developers to understand the structure and where to find components easily. Members of the team can easily test out components with different properties before even using them in the real project using Storybook.

In the next section, we will explore how to create an internationalized application in Vue.js 3, which allows your application to not be limited to a single language and cultural setting.

Implementing internationalization and localization

The essence of building an enterprise application is to support local markets around the world and to achieve this, that's where internationalization comes into place.

The process of preparing software to support local languages and the cultural settings of other geographical locations is called **internationalization (I18n)**.

I18n is often misrepresented as **localization (L10n)** and sometimes even translation.

While Il8n is a product development approach that is focused on making one code base capable of supporting worldwide languages and locale-specific formatting and behaviors, L10n makes a product specific to a target market or region(s), including translation of the interface and possible adaptation of terminology and more.

In this section, we will first look into the benefits of internationalizing the software and further explore how to implement I18n in Vue 3.

Benefits of internationalizing software

The benefits of creating internalized software are enormous and some of these are listed here:

- It creates higher-quality software that meets the technical and cultural needs of multiple locales
- It provides greater in-country customer acceptance and satisfaction
- It provides a single source code for all languages of the product
- Internalized software reduces time, cost, and effort for L10n
- Internalized software is simpler, and supports easier maintenance for future iterations of the product

Market acceptance is one of the major problems that arises when software is not fully internalized before or after release.

Therefore, we will look at how to implement I18n in our enterprise Vue 3 application right from the development phase.

Installing Vue I18n

In Vue 3, Vue-I18n is a great compatible plugin that is used to implement I18n, and it easily integrates some localization features into your Vue.js application.

Follow these steps to internationalize your app:

1. There are different ways to install the package according to the official documentation (https://vue-i18n.intlify.dev/installation.html), but we will install it using the npm command, as shown here:

    npm install vue-i18n@9

2. After installation, inside the Vue 3 main.js file, add the following script:

    ```
    import { createApp } from 'vue'
    import { createI18n } from 'vue-i18n'

    const i18n = createI18n({
      // something vue-i18n options here ...
    })
    const app = createApp(App)
    app.use(i18n)
    app.mount('#app')
    ```

    With the preceding setup, you should have internalization added to your Vue project, but it will easily get bloated when developers start adding translations. So, we recommend creating a locales folder where every locale-related configuration will be added.

3. Let's create the folder and the files inside the root directory like so:

    ```
    mkdir src/locales
    touch src/locales/index.js src/locales/en.json src/locales/fr.json src/locales/de.json
    ```

4. Next, inside each of the translation files, add the following codes and other translations:

    ```
    {
      "welcomeMsg": "Welcome to Your Vue.js App",
        ....
    }
    ```

5.   Inside the `index.js` file, add the following scripts to import different locales:

```
import en from «./en.json»;
import fr from "./fr.json";
import de from «./de.json»;

const messages = {
 en,
 fr,
 de,
};

export default messages;
```

6.   Lastly, add the files to your `createI18n` configuration in your `main.js` file:

```
import locales from "./locales/index.js";

const i18n = createI18n({
 locale: "en", // set locale
 fallbackLocale: "en", // set fallback locale
 messages: locales, // set locale messages
});
```

Arranging your files and folder in this structure allows for easy adoption and maintainability. Let's look at the final structure of our project, including internationalization, in the following screenshot:

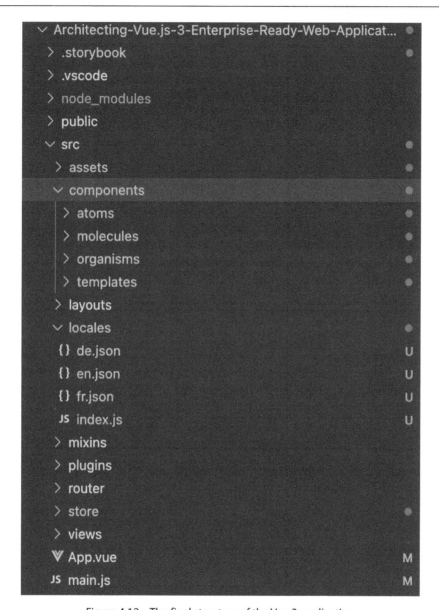

Figure 4.12 – The final structure of the Vue 3 application

Summary

This chapter reviewed in more depth architecting large-scale web apps with Vue 3. We discussed the structure and file architecture by diving deeper into the law of predictability and how to use community-recommended packages to inform predictability in your Vue.js 3 enterprise-ready application.

We also covered in detail how to use micro frontend architecture to your advantage. Also, we discussed how to implement an Atomic Design with Storybook to streamline your component directory and make your enterprise project less difficult to understand.

Next, we discussed how to add I18n to your Vue application. We also discussed the benefits and how to properly integrate I18n into your Vue 3 application.

In the next chapter, we will explore GraphQL, GraphQL Apollo Server 2, queries, mutations, and how to integrate these technologies into your Vue.js 3 application. In addition, you will learn how to utilize GraphQL to deliver scalable and high-performing applications.

Part 3: Vue.js 3 Enterprise Tools

In this part, you will learn about DevOps and Docker. You will containerize your web app and deploy a container to Google Cloud Run using CLI tools. Then, you will leverage advanced CI techniques to build a container-based CI environment, leveraging a multi-stage Dockerfile.

This part will also explore GraphQL and how it can be implemented and integrated with Vue.js 3 to deliver an enterprise-ready web application. In addition, we will build an enterprise Pinterest clone to demonstrate our knowledge of GraphQL at an enterprise level.

This part comprises the following chapters:

- *Chapter 5, An Introduction to GraphQL, Queries, Mutations, and RESTful API*
- *Chapter 6, Building a Complete Pinterest Clone with GraphQL*
- *Chapter 7, Dockerizing a Vue 3 App*

5

An Introduction to GraphQL, Queries, Mutations, and RESTful APIs

In the previous chapters, we explored different libraries and methods to develop large-scale enterprise applications using Vue 3. In this chapter, we will first understand what GraphQL is and how it is different from REST. Next, we will explore GraphQL, GraphQL Apollo Server 2, queries, and mutations, and how to integrate these technologies into your Vue.js 3 application. In addition, you will learn how to utilize GraphQL to deliver scalable and high-performing applications.

We will cover the following key topics in this chapter:

- An introduction to GraphQL
- Understanding queries and mutations
- Integrating GraphQL Apollo Client with Vue 3

Also, in this chapter, you will learn how to integrate GraphQL into Vue 3 and structure it properly following the law of predictability by implementing a login and register authentication system using the GraphQL Apollo client and Vue 3.

Technical requirements

To get started with this chapter, we recommend you read through *Chapter 4*, *Architecture for Large-Scale Web Applications*, where we explored building large-scale enterprise applications with different industry-standard structuring, architecture, and standards.

All the code files for this chapter can be found at: `https://github.com/PacktPublishing/Architecting-Vue.js-3-Enterprise-Ready-Web-Applications/tree/chapter-5`.

An introduction to GraphQL

GraphQL is the new buzzword in the API development industry. While REST remains the most popular way to expose data from a server, it comes with many limitations that GraphQL tends to solve.

GraphQL is a query language created and maintained by Facebook. The purpose of creating GraphQL is to build client applications based on intuitive and flexible syntax for describing their data requirements and interactions.

One of the benefits of GraphQL is that we have a single endpoint to access all data from the server instead of having multiple endpoints in REST.

In this section, we will explore everything you need to know about GraphQL, the different unique features of GraphQL, and why you should consider GraphQL instead of the RESTful API design pattern. Lastly, we will work you through creating and setting up your first GraphQL server with Express.

What is GraphQL?

As per the official documentation (`https://graphql.org/`),

> *GraphQL is a query language for APIs and a runtime for fulfilling those queries with your existing data. GraphQL provides a complete and understandable description of the data in your API, gives clients the power to ask for exactly what they need and nothing more, makes it easier to evolve APIs over time, and enables powerful developer tools.*

GraphQL is a server-side runtime for executing queries using the type system you define for your data. Also, GraphQL is not tied to any specific database or storage engine but instead backed by your existing code and data.

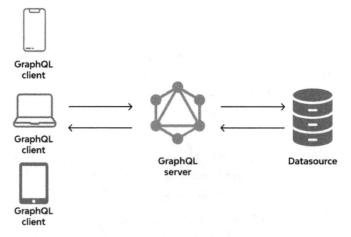

Figure 5.1 – A diagram explaining GraphQL (Source: https://www.wallarm.com/what/what-is-graphql-definition-with-example)

The GraphQL type system defines various data types that can be used in a GraphQL application. This type system helps to define the schemas that will be used in the GraphQL application.

To create a GraphQL service, you need to start by defining schema types and creating fields on those types, then providing functions to be executed on each field.

For example, we can define a new schema type called `Photo` in the following code snippet to demonstrate how types work in GraphQL:

```
type Photo {
  id: ID!
  name: String!
 url: String!
 description: String
 }
```

Now we have an idea of GraphQL and have seen how to define a GraphQL schema type. Next, let's explore the features of GraphQL before we dive deeper into creating GraphQL queries and resolvers.

Features of GraphQL

GraphQL comes with excellent features. We are going to explore a few of the features of GraphQL in the following subsections.

Easy to get started

The learning curve of GraphQL is easy, especially for developers who are familiar with building APIs with RESTful design patterns.

Users can get started with GraphQL with small queries for fetching data right away and learn about the advanced features a bit later.

Built for interactive apps

GraphQL is built for real-time and interactive applications because changes between the client and the server happen almost immediately, giving a swift response.

Small and flexible

GraphQL allows users to request and receive the exact data that was requested. This feature solves the problem of over- and under-fetching with RESTful APIs.

Universally compatible

GraphQL Apollo is built to be compatible with any build setup, any GraphQL server, and any GraphQL schema.

Incrementally adoptable

GraphQL is built to make integrating into a new or existing project effortless without breaking any changes. It is easily adaptable.

Why use GraphQL instead of REST?

In this section, we will identify some properties of GraphQL and discuss why you should use GraphQL for your subsequent API development instead of RESTful APIs. We will discuss a few of these properties. Additionally, we will get into a detailed comparison of these technologies with the GraphQL versus RESTful API guide in this section.

The following subsections mention the top five reasons why you should use GraphQL instead of RESTful API.

Strongly-typed schema

In GraphQL, **Schema Definition Language** (**SDL**) is used to define the contract between the client and the server and to define how the client accesses the data in the server. GraphQL uses a strong type system to define the capabilities of the API. All the APIs exposed to the client are written down in a schema called the SDL.

Once these schemas are defined, both the frontend and the backend can communicate separately without any further changes or assistance since the frontend knows that the data in the schema is always going to be in sync or consistent across the system.

This solves the data inconsistency problem in REST. The frontend expects a specific dataset but retrieves a different one due to changes since no consistent schema is defined.

No over-fetching or under-fetching

The issue of over- or under-fetching is a known problem with RESTful APIs where clients download data by hitting endpoints that return fixed data structures or retrieve either excessive amounts of data or less than what they expected.

Over-fetching is the problem of fetching more data than what is required in the app. It can also mean fetching more data than what is required to fulfill the request. With a RESTful API, you have to fetch all the user's details or create a new endpoint that returns only the names of all the users of your application just to display only the name of the user on your frontend application. While in GraphQL, you can use just a single query to return only the name of all the users or any other details by creating a separate query or endpoint.

Under-fetching is rare, but it happens when the specific endpoint does not provide all the required information. The client must make additional requests to access the other information as needed.

GraphQL solves this problem by providing a medium for the client to specify the information needed, and it returns exactly the required information.

Saving time and bandwidth

The problem of over-fetching can result in higher bandwidth consumption for clients, which may, in time, cause lagging in your application. Using RESTful API design patterns, it is more time-consuming to sort out the information needed from an enormous payload.

GraphQL allows you only to select what you need, thereby reducing the amount of payload transferred over the network.

Multiple endpoints

One of the significant problems of RESTful APIs is having too many endpoints to access information. For instance, if a client wants to access a particular user by their ID, you will be presented with /users/1. Also, if you're going to access that user's photos, you will have to send a request to another endpoint, /users/1/photos.

In GraphQL, you have a single endpoint, and you don't need to send multiple requests to retrieve different information about the user.

With GraphQL, you can access all the user's information in a single request, as shown here:

```
{
    me {
        name,
        photos {
            title,
url
        }
    }
}
```

The preceding script will only access name of the user and title and url of all the user's photos.

Versioning is not required

Versioning is a feature of RESTful APIs where different versions are assigned to an API when changes and updates are made to it. This is done to avoid breaking changes in production that might cause issues for the user.

If we want users to use our new features in the latest version, we have to force them to update the application, which is not a good user experience.

In GraphQL, there is no need to worry about versioning and forcing users to update their application to use the new changes since it happens automatically, and clients only implement the features available in the SDL.

Here, we have discussed the different features and benefits of using GraphQL and why you should consider using GraphQL instead of RESTful APIs. In the next section, we will further discuss the difference between GraphQL and RESTful APIs.

The difference between GraphQL and RESTful APIs

The core difference between GraphQL and REST is that GraphQL is a specification and a query language, while REST APIs are an architectural concept for network-based applications.

Both concepts play an important role in creating and maintaining scalable microservices and large-scale enterprise applications.

Therefore, choosing a particular technology to go with will depend on your level of expatriation in each technology, which one your team is comfortable with, and which one makes your product development easier and faster.

The following points show you the other differences you might want to consider:

- GraphQL is a query language used to solve problems by integrating APIs, while REST API is an architectural style that describes the conventional standard for designing APIs

- Additionally, with GraphQL, you can use a single endpoint to access all the data in your server without the need for multiple endpoints, while REST API allows multiple endpoints and a set of URLs that each exposes a single resource, which can be confusing to understand

- GraphQL uses a client-driven architecture and lacks an in-built caching mechanism, while REST API uses a server-driven architecture and uses caching automatically

- No API versioning is required in GraphQL, and its response is only in JSON format, while REST APIs support multiple API versioning and allow response output in XML, JSON, or YAML

- GraphQL offers type safety and auto-generated documentation and it also allows schema stitching and remote data fetching, while REST API doesn't offer type safety or auto-generated documentation, and simplifying work with multiple endpoints requires expensive custom middleware

- GraphQL is also an application-layer server-side technology used for executing queries with existing data, while REST is a software architectural style used to define a set of constraints for creating web services

- GraphQL can be organized in terms of a schema, while REST is not arranged or organized in schemas but is arranged in terms of endpoints

- The development speed in GraphQL is faster when compared with REST APIs

- The message format for GraphQL mutations should be a string, while the message format for REST APIs can be anything

- GraphQL uses metadata for query validation, while REST does not have cacheable machine-readable metadata

We have explored the difference between GraphQL and REST API to give you an insight into which is better for your enterprise application. We will allow you to make a choice, but we will explore GraphQL in more depth in the following sections. In the next section, we will discuss queries and mutations, expanding more on how to create your first query and mutation.

Understanding queries and mutations in GraphQL

Queries and mutations are vital in GraphQL because they are the only way to access or send data to the GraphQL server from your frontend.

Using queries

GraphQL queries define all the queries that a client can run on the GraphQL API. If you're familiar with REST APIs, it is synonymous with the popular GET requests.

You can define GraphQL queries in many ways, but defining a root query to wrap all your queries is recommended.

The following code snippet shows you how to define a root query called RootQuery:

```
type RootQuery {
  user(id: ID): User     # Corresponds to GET /api/users/:id
  users: [User]          # Corresponds to GET /api/users
  photo(id: ID!): Photo #Corresponds to GET/api/photos/:id
  photos: [Photo]         # Corresponds to GET /api/photos
}
```

You can also define individual queries for User and Photo like so:

```
type User {
id: ID!
name:String!
email: String!
```

```
}

type Photo {
id: ID!
title:String!
description: String
url: String!
user: User
}
```

With the preceding queries, we have successfully defined endpoints to retrieve users and photos corresponding to how it can be done with REST API using the GET method.

Next, we will explore how to create or update data on our GraphQL API using mutations.

Using mutations

Mutations in GraphQL are used to create, update, and delete data from our GraphQL API, and are synonymous with REST API's POST, PUT, and DELETE methods, respectively.

Creating mutations in GraphQL is simple; take a look at the following snippet:

```
type RootMutation {
  createUser(input: UserInput!): User
  # Corresponds to POST /api/users
  updateUser(id: ID!, input: UserInput!): User
  # Corresponds to PATCH /api/users
  removeUser(id: ID!): User
  # Corresponds to DELETE /api/users

  createPhoto(input: PhotoInput!): Photo
  updatePhoto(id: ID!, input: PhotoInput!): Photo
  removePhoto(id: ID!): Photo
}
```

The preceding snippet shows how to create mutations. Furthermore, we have created createUser, updateUser, and removeUser to create, update, and delete users from the GraphQL API.

Most importantly, for GraphQL to connect to our database and carry out the operations in both queries and mutations, we need to define a resolver, which we will cover in the next subsection.

Resolvers

A GraphQL resolver connects our queries and mutations to the right methods for execution. It informs GraphQL what to do when each of these queries/mutations is requested. It is a basic function that does the hard work of hitting the database layer to do the CRUD operations, hitting an internal REST endpoint, or calling a microservice to fulfill the client's request.

Let's map the queries/mutations we have created in the preceding sections to a resolver that will be called when any of the queries/mutations are requested:

```
import sequelize from '../models';

export default function resolvers () {
  const models = sequelize.models;

  return {

// Resolvers for Queries
    RootQuery: {
      user (root, { id }, context) {
        return models.User.findById(id, context);
      },
      users (root, args, context) {
        return models.User.findAll({}, context);
      }
    },

// Resolvers for Mutations
RootMutation: {
  createUser (root, { input }, context) {
    return models.User.create(input, context);
  },
  updateUser (root, { id, input }, context) {
    return models.User.update(input, { ...context,
                                    where: { id } });
  },
  removeUser (root, { id }, context) {
    return models.User.destroy(input, { ...context,
```

```
                                            where: { id } });
    },
    // ... Resolvers for Photos
  }
};
}
```

To foster understanding, we imported our model from `sequelize`, which is a database **object-relational mapping (ORM)** to manipulate database tables with defined methods.

Next, we created and exported a `resolver` function, which returns an object containing `RootQuery` and `RootMutation`.

Next, inside the `RootQuery` and `RootMutation` objects, we resolve each of the methods with the appropriate business logic to be executed.

For example, when a client requests all users, the GraphQL frontend client will call the user's queries defined in the **Queries** section, and the GraphQL engine will call the user's resolver to retrieve all the users using the `sequelize` ORM. The same logic applies to all the mutations.

In this section, we explained of how GraphQL works and how you can create your own GraphQL API to understand the best way to structure your GraphQL client in the frontend for enterprise projects.

In the next section, we are going to explore the best way to structure your enterprise Vue.js application with GraphQL for scalability and faster team adoption. Remember the law of predictability for your teams that we discussed in *Chapter 3, Scaling Performance in Vue.js 3*.

Integrating GraphQL Apollo Client with Vue 3

It is tempting to ask what the best way to integrate GraphQL client into Vue 3 is and how to structure it in an enterprise project to foster faster adoption by team members, including new team members.

In my career as a full-stack software engineer, I recently joined a fintech company using Vue 3 and GraphQL to disrupt the fintech industry in Germany, and I was so impressed at the arrangement of such a large code base and how easy it was for me to jump right into solving my first task.

There are many ways to structure your Vue 3 project with GraphQL, but I want to outline the best way I have seen that works for small- or large-scale enterprise projects, including the fintech one I worked on.

GraphQL Apollo Client is a JavaScript library used to connect to the GraphQL server to interchange data. With the library, you can connect to the server, send requests, and receive responses from the server.

First, we will start by listing and installing the necessary packages for GraphQL and the GraphQL client in Vue 3.

Installation

Follow these steps to install all the necessary packages:

1. Type in the following commands to install `graphql`, `graphql-tag`, `apollo-composable`, and `apollo client`. These are the recommended libraries from the official documentation used to communicate with the GraphQL server using Vue 3:

    ```
    npm install --save graphql graphql-tag @apollo/client @
    vue/apollo-composable
    ```

2. After installation, we will create a new file called `apollo.config.js` inside our `plugins` folder from the folder structure we created in *Chapter 3, Scaling Performance in Vue.js 3* and add the following scripts to configure the `graphql` client:

    ```js
    ```js
    import { ApolloClient, createHttpLink, InMemoryCache }
    from '@apollo/client/core'

    // HTTP connection to the API
    const httpLink = createHttpLink({
      // You should use an absolute URL here
      uri: 'http://localhost:3020/graphql',
    })
    // Cache implementation
    const cache = new InMemoryCache()
    // Create the apollo client
    const apolloClient = new ApolloClient({
      link: httpLink,
      cache,
    })
    export default apolloClient
    ```
    ```

3. Lastly, inside your Vue 3 `main.js` file, add the following script to the existing one:

    ```js
    ```js
    import { createApp, provide, h } from 'vue'
    import apolloClient from "./plugins/apollo.config";
    import { DefaultApolloClient } from '@vue/apollo-
    composable'
    ```
    ```

```
...
const app = createApp({
  setup () {
    provide(DefaultApolloClient, apolloClient)
  },
  render: () => h(App),
})
...
```

In the preceding steps, we demonstrated how to install the GraphQL client and completely set it up with the other libraries in our Vue 3 enterprise application. In the next section, we will discuss the best practices for structuring our Vue 3 application with GraphQL.

Structuring GraphQL

After successfully installing and configuring Apollo Client with Vue 3, let's structure our queries and mutations around the law of predictability to enable old and new team members to easily adapt to the project.

Create a new folder inside the src folder called graphql. This new folder will contain all our queries and mutations grouped in to different directories according to the features of the application.

Let's take an example from the schema we developed in the previous section. From the schema, it is clear that our project has User and Photo features, so we will create different folders inside the general graphql folder for these specific features.

Create new folders using the following command line or manually from your code editor:

```
mkdir src/graphql
mkdir graphql/users graphql/photos
```

You should have a new folder structure, as shown in the following screenshot:

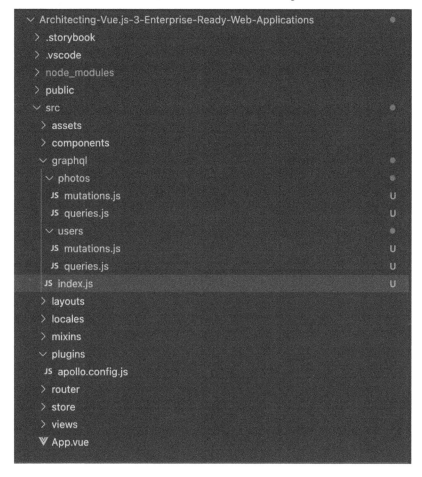

Figure 5.2 – The folder structure after installing and setting up GraphQL

After installing and setting up GraphQL, your folder structure should look like the preceding screenshot, containing each feature with its corresponding queries and mutations file.

Structuring your GraphQL API this way makes it easy for your team to automatically understand where to find anything related to GraphQL queries or mutations and in which feature they are looking for them.

Now that we have our folder structure figured out for predictability, in the next section, we will demonstrate with a practical exercise using the structure we learned about in the previous sections to authenticate a user into our application using GraphQL Apollo Client and JWT installed on our GraphQL server.

JWT authentication for sign in/sign up

With the structure we have on the ground, it becomes easy to add new features and write our queries/mutations without scattering them in the code base.

Let's demonstrate how to implement the login and register authentication process with GraphQL and Vue 3 by following these steps:

1. Firstly, create a new folder called `auth` inside the `graphql` folder and add `mutations.js` inside it:

```
mkdir graphql/auth
touch graphql/auth/mutations.js
```

2. Inside the newly created mutation file, add the following script for registration and login endpoints:

```
import { gql } from "graphql-tag";

export const LOGIN_USER = gql`
 mutation loginUser($input: LoginUserInput!) {
   loginUser(data: $input) {
     id
     name
     email
     token
   }
 }
`;


export const REGISTER = gql`
 mutation register($input: RegisterUserInput!) {
   register(data: $input) {
     id
     name
     email
   }
 }
`;
```

3. Next, export the mutation inside the `graphql/index.js` file we created earlier to make it available throughout our Vue application:

```
export * from "./auth/mutations";
```

The `export` script makes importing our GraphQL queries and mutations a lot easier.

Next, we will look at how to call the authentication mutation inside the Vue 3 component and retrieve the user's information.

With the `vue-composable` library we installed earlier, we can use different GraphQL hooks, such as `useMutation` and `useQuery`, to manipulate the GraphQL API.

4. Create a `login` component inside the `organisms` folder and add the following code:

```
<template>
  <form @submit.prevent="login">
    <TextField
      v-model="email"
      required
      type="email"
      class="mb-6"
      name="email"
      label="Email Address"
      placeholder="Enter your email address"
    />
      <TextField
        v-model="password"
        required
        minlength="6"
        class="mb-6"
        type="password"
        maxlength="50"
        placeholder="Enter your full password"
        label="Password"
      ></TextField>
      <div class="flex justify-between mb-6">
        <CheckField id="remember" v-model="remember">
          Remember me</CheckField>
```

```
      </div>
      <div class="flex justify-center my-9 w-full">
        <Button>Sign In</Button>
      </div>
    </form>
  </template>
```

The first code snippet shows the template and the view part of the component; it has a form component with four child components, including TextField, CheckField, and Button:

```
<script>
import { LOGIN_USER } from "../../graphql";

export default {
 setup(props) {
    const email = ref("");
    const password = ref("");
    const remember = ref(false);

    const { mutate: loginUser } =
      useMutation(LOGIN_USER, () => ({
      variables: {
        email: email,
        password: password,
        remember,
      },
    }));

    const login = () => {
      const user = loginUser();
      if (!user) return

     // Save State and Redirect to Dashboard
    };

    return {
```

```
        login,
        email,
        password,
        remember,
      };
    },
  };
</script>
```

The script section of the component performs the business logic; it has many properties and a function called Login, which performs the authentication process for your application.

The preceding snippet shows how to implement the login functionality for our project. You can implement the registration component or take a look at the repository (https://github.com/PacktPublishing/Architecting-Vue.js-3-Enterprise-Ready-Web-Applications/tree/chapter-5) for the complete code base for this chapter.

Figure 5.3 – A screenshot showing the implementation of a login form

Summary

This chapter delved deeper into GraphQL, GraphQL Apollo Server 2, queries, mutations, and how to integrate these technologies into your Vue.js 3 application. In addition, we learned how to utilize GraphQL to deliver scalable and high-performing applications.

We also covered in detail how to integrate GraphQL into Vue 3 and properly structure it following the law of predictability.

Lastly, we demonstrated how to implement a login and register authentication system using GraphQL Apollo Client and Vue 3.

In the next chapter, you will learn how to build a complete Pinterest clone with Vue 3 and GraphQL. You will utilize your knowledge of GraphQL to develop and deliver an enterprise application such as Pinterest using Vue 3 and GraphQL.

6
Building a Complete Pinterest Clone with GraphQL

In the previous chapter, we explored GraphQL, GraphQL Apollo Server 2, queries, mutations, and how to integrate these technologies into your Vue.js 3 application. In addition, we learned how to utilize GraphQL to deliver scalable and high-performing applications. This chapter will demonstrate how to build a complete Pinterest clone with Vue 3 and GraphQL. You will utilize your knowledge of GraphQL to develop and deliver an enterprise application such as Pinterest using Vue 3 and GraphQL. Furthermore, you will learn how to create and manage your backend APIs using a popular headless **content management system (CMS)** called Strapi.

We will cover the following key topics in this chapter:

- An introduction to Strapi
- Scaffolding a Strapi project
- Building the collections
- Building a Vue 3 Pinterest app
- Connecting the frontend and backend
- Testing the app

By the end of this chapter, you will have learned how to create a scalable and high-performing Pinterest clone application with GraphQL.

Technical requirements

To start with this chapter, I recommend you read through *Chapter 5, An Introduction to GraphQL, Queries, Mutations, and RESTful APIs*, where we explored GraphQL, GraphQL Apollo Server 2, queries, mutations, and how to integrate these technologies into your Vue.js 3 application.

All the code files required for this chapter can be found at `https://github.com/ PacktPublishing/Architecting-Vue.js-3-Enterprise-Ready-Web- Applications/tree/chapter-6`.

An introduction to Strapi

Strapi is an open-source headless CMS based on Node.js that is used to create and manage different forms of content using a RESTful API and GraphQL.

Additionally, Strapi makes developing, deploying, and maintaining APIs faster and can be configured to consume content via APIs using any HTTP client or GraphQL-enabled frontend.

These benefits are the reason we will use Strapi to create the backend of our Pinterest clone application, so we can focus more on the frontend without having to pay much attention to scaling the backend.

In the next section, we will work through scaffolding a Strapi project, building out all the collections we need in our project, and seeding out the Strapi account with dummy data.

Scaffolding a Strapi project

Starting a new Strapi backend project is very easy and works precisely like installing a new framework using the CLI tool.

We will scaffold a full-blown backend application by running any of these simple commands and testing it in our default browser:

```bash
npx create-Strapi-app strapi-pinterest-api --quickstart
    # OR
yarn create straps-app strapi-pinterest-api --quickstart
```

The preceding command scaffolds a new Strapi API into the specified folder. Next, run the following command to build and deploy your newly created backend API with Strapi:

```bash
npm run build
```

```
npm run deploy
```

These preceding two commands should build your app and deploy it so you can easily test it out by typing the following URL (`localhost:`) in your default browser if it doesn't open automatically.

Most importantly, the last command will also open a new tab with a page to register your new admin user of the system. Go ahead and fill out the form and click on the **LET'S START** button to create a new admin user.

Here is what the page looks like:

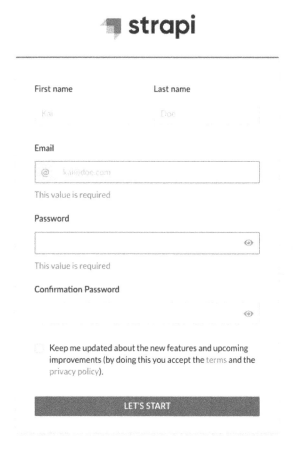

Figure 6.1 – The Strapi registration page

After signing up as the admin of the platform, you will have admin access to create different collections and set up different users and permission levels for your API.

In the next section, we will explore how to create different collections that will correspond to our Pinterest API resources. We will also set up different permission levels to limit what each user in our Pinterest application can access.

Building the collections

Strapi uses collections to denote resources; for example, if your application is a news application and you want to create a backend that will process posts, comments, and so on, you will create it as a Posts, Comments collection in Strapi.

However, since we are building a Pinterest clone, we will create the following collections: **Photo(Pin)**, **Board**, and **User**, and each of these collections will contain their respective fields, as demonstrated in the following steps.

To demonstrate, we will create a simple Photo(Pin) collection that will store the details of a specific photo in our app. In Pinterest, it is called **PIN**, but we will prefer to call it as **PHOTO** since we started with that in the previous chapters.

Now to store the details of our photo, we will create a new **Collection Type** called `photos` in the Strapi dashboard.

The `photos` collection will have the following fields: **title**, **url**, **user_id**, and **description**. These fields are imaginative and can change as we process them into the book:

1. To create our first **Collection Type**, log into the **Admin** dashboard and click **Collection Type Builder** on the left side of the page. Next, click on **Create New Collection Type** still on the left side of the page and fill in `photos` as the display name.

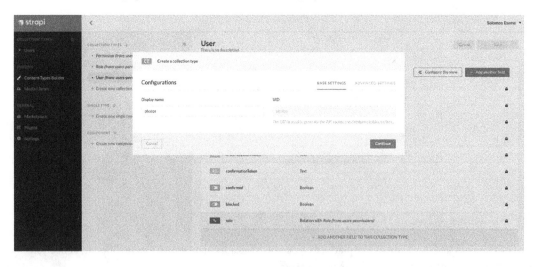

Figure 6.2 – The Strapi collection dashboard

2. Next, choose the data type for your field, as shown in the following screenshot:

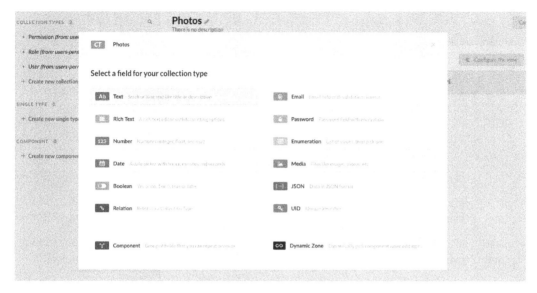

Figure 6.3 – The Strapi Photos collection

3. Next, enter the name of the field for your **Photos** collection and click on + **Add another field** to enter another field:

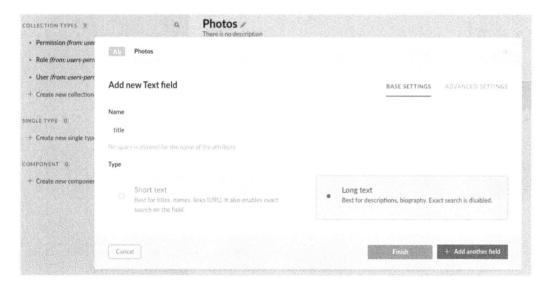

Figure 6.4 – Strapi | Add new Text field

4.   Repeat the process until you exhaust the list of fields for your collection and click on **Finish**.

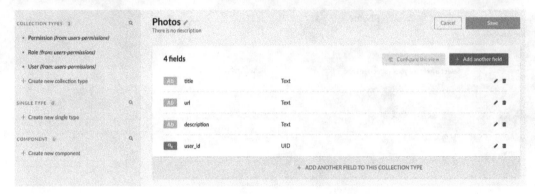

Figure 6.5 – The Strapi Photos dashboard

5.   Lastly, click on **Save** and repeat the process for **Users**, **Boards**, and other collections you might want to add. You should have all the collections listed, as shown here:

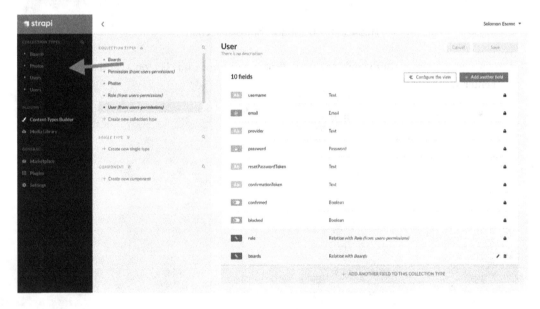

Figure 6.6 – Strapi | User fields

> **Important note**
>
> You can learn more about Strapi collections here – `https://docs.strapi.io/user-docs/latest/content-types-builder/introduction-to-content-types-builder.html` – or watch this video – `https://www.youtube.com/watch?v=bStlyMB0NEw` – to see how we created all the collections we will use in the project.

In the next section, we will explore how to seed fake data into the Strapi collection to enable us to display some photos before we start adding new ones.

Seeding data

After successfully creating the collections, we will seed some data so that we have plenty of photos, boards, and users to work with.

Take the following steps to seed some data into the collections we have created. First, we will seed some photo information, including the photos, and create a board that will house some of these photos and a user who is taking these actions:

1. To seed dummy data for our Pinterest project, we will click on each of the collections we have created and click + **Add New Users**, as shown in this screenshot:

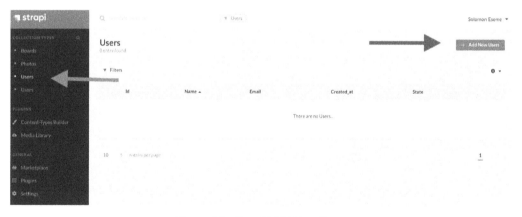

Figure 6.7 – Strapi | Add New Users

2. Next, fill in the information needed to create a single user of our Pinterest application and click on **Save**, as shown in this screenshot:

Figure 6.8 – Strapi | Create an entry

3. Click on **Publish**, and you should have your new user added to the **Users** collection, as seen here:

Figure 6.9 – List all users

Create more fake data by repeating the process for all other collections, such as **Photos** and **Users**, for testing. Later in the next section, we will learn how to programmatically create data in the Strapi collection and build our Pinterest clone using Vue 3.

Building a Vue 3 Pinterest app

In the previous section, we explored creating the backend of our Pinterest application using Strapi. In this section, we will create the frontend using Vue 3.

However, it is important to note that since this is a demo, we will only abstract the slightest part of Pinterest to represent the application. Developing the full Pinterest application will require effort, teams, and resources.

We will continue by using the official project we created for this book. In the previous chapters, we added internationalization, structured the project, and built out the login form, and we will continue by including other necessary files to make up a full-blown Pinterest clone application.

Most importantly, I will be using Tailwind CSS as my CSS framework for this project, and since it's beyond the scope of this book, you can visit the official documentation to set it up with Vue 3.

You can clone the project from this repository – `https://github.com/PacktPublishing/ Architecting-Vue.js-3-Enterprise-Ready-Web-Applications` – to jump right in.

Here is a demo of what we are building:

Figure 6.10 – Pinterest preview

We have everything separated properly and structured in a scalable format for our application. The following shows how we structure the `HomeOverview` component that represents the home page:

```
<template>
 <main>
   <Header></Header>
   <section style="mt-20">
     <Cards />
   </section>
 </main>
</template>

<script setup>
import Header from '../organisms/Header.vue'
import Cards from '../organisms/Cards.vue'
</script>
```

It contains the `Header` and `Cards` components. We could use `alias` to import these components making the import URL shorter, but I preferred to show you the relative path.

In the next section, we will start building out the frontend project by creating the `Cards` component and implementing the logic to display all the photos we have created previously.

Generating the Cards component

Let us take a look at the `Cards` component first to explore the content in it. The `Cards` component houses the logic behind displaying the photo we have created and stored in our Strapi instance, and you can see this in the following code snippet:

```
<template>
 <div class="pin_container sm:justify-center">
   <div class="card card_small h-[16.25rem] relative">
     <!-- // Medium -->
     <Card />
   </div>

   <div class="card card_medium h-[20.625rem] relative">
     <!-- // Small -->
     <Card />
```

```
    </div>

    <div class="card card_large h-[28.125rem] relative">
      <!-- // Smaller -->
      <Card />
    </div>

    <div class="card card_smaller h-[11.063rem] relative">
      <!-- // Medium -->
      <Card />
    </div>

    <div class="card card_small h-[16.25rem] relative">
      <!-- // Large -->
      <Card />
    </div>
  </div>
</template>

<script setup>
import Card from '../molecules/Card.vue';
</script>
```

The `Cards` component is where the magic happens, as it represents the collections of each Photo (Pin) in the application.

Here is a preview of the code:

Figure 6.11 – Cards preview

Firstly, we display the cards based on their sizes (smaller, small, medium, and large). This helps us to get the same previews as Pinterest.

You can clone the full frontend code and the Strapi backend from the `chapter 6` branch in the repository here: `https://github.com/PacktPublishing/Architecting-Vue.js-3-Enterprise-Ready-Web-Applications/tree/chapter-6`.

In the previous section, we demonstrated how to build a simple Pinterest clone by creating different `Card` components to represent the Pin information. In the next section, we will learn how to connect our Strapi backend to the Pinterest frontend we have created using Vue 3.

Connecting the frontend and backend

The most interesting part is how we structure our API requests to accommodate maintainability and easy adaptability, following the best practices we have learned from the previous chapters.

The following screenshot shows our folder structure containing the GraphQL endpoints according to the features we currently have in our application and in the Strapi backend we developed for this project:

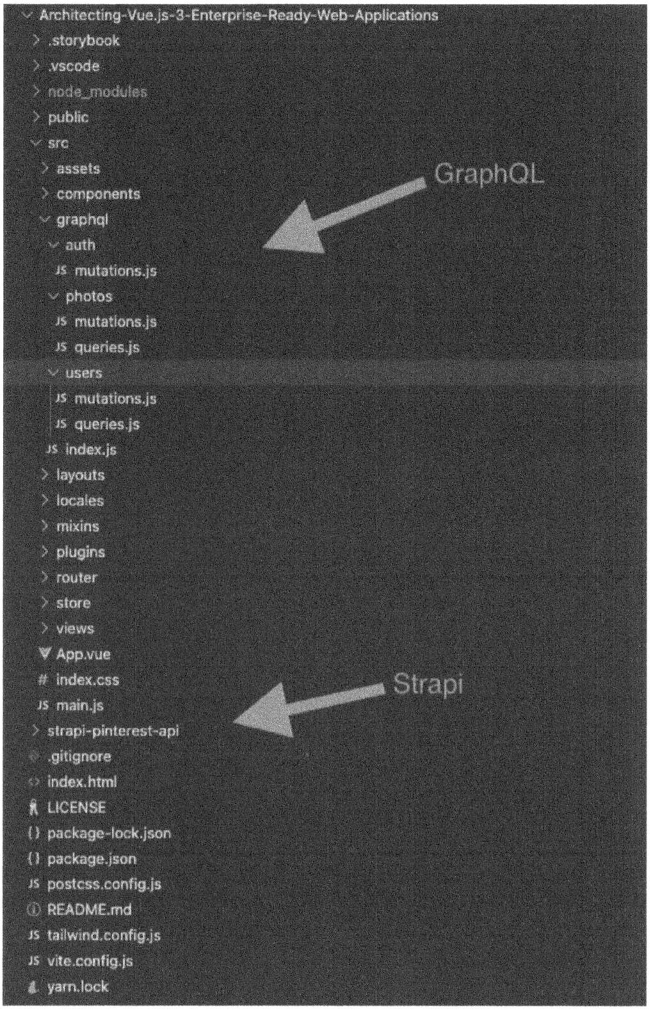

Figure 6.12 – A screenshot of the new folder structure with GraphQL and Strapi

In the graphql folder, we have defined three folders, namely auth, photos, and users, which represent the features our current project will have. Let's look at what these folders contain.

The auth folder

The auth folder contains only a single mutation which will handle all the authentication and authorization functionalities. It will contain mutations, such as register, login, forgotPassword, sendForgotPasswordEmail, and so on.

The photos folder

The photos folder is the most complex because it contains all the functionalities of the Pinterest application demo. It contains both mutations and queries, which is why we have created different files for it.

Some of the mutations are createPin, createBoard, updatePin, updateBoard, deletePin, and so on. These mutations send different actions to our Strapi backend server to perform different actions.

Additionally, we have the queries.js file, which contains all the queries to retrieve the different types of data from our Strapi backend server.

Some of the queries found in this file include getPin, getBoard, getBoards, getPins, getUserPins, getUserBoards, and so on.

The users folder

The users folder contains all the user-related functionalities of the Pinterest application demo. It contains both mutations and queries.

Some of the mutations are createUser, updateUser, deleteUser, and so on. These mutations send different actions to our Strapi backend server to perform different user-related actions.

Additionally, we have the queries.js file, which contains all the queries to retrieve the different types of user-related information from our Strapi backend server.

Some of the queries found in this file include getUser, getUsers, and so on.

Implementing a login example with GraphQL

The code snippet for each of the methods inside each of the mutations and queries can be found on the official GitHub repository for the respective chapter.

However, the following is an example of how to log in to our application using the GraphQL mutation we have defined inside the auth folder.

First, we import the respective mutation we want to use inside any component or page, as shown here:

```js

import { LOGIN_USER } from '../../graphql';
```

Because of the index.js file inside the graphql folder, the folder location is shortened a little; you can reduce it more depending on your use case.

This is why it is important to always add an export inside the index.js file for any GraphQL mutation or query created.

The following steps show you how to implement logging authentication with GraphQL in our Pinterest clone application example.

Step 1

When a user tries to log in, we execute the `loginUser` function to retrieve some user-specific data:

```
const login = () => {
 const user = loginUser();
 if (user) {
   // Save State and Redirect to Dashboard
   logged.value = true;
 }
};
```

Step 2

The `loginUser` function executes the `LOGIN_USER` mutation we imported using the `useMutation` hook imported from the Apollo Composable library as follows:

```
import { useMutation } from '@vue/apollo-composable';

const { mutate: loginUser } = useMutation(LOGIN_USER, () => ({
 variables: {
   email: email,
   password: password,
   remember,
 },
}));
```

Code walkthrough

If you haven't used GraphQL with Vue.js before, here is a quick walkthrough.

`useMutation` executes any mutation using the information passed in the `variables` object:

```
variables: {
   email: email,
   password: password,
   remember,
 },
```

Every useMutation execution returns the mutate function, which we rename to the name of the executed mutation called when our users try to log in. The loginUser function executes the LOGIN_USER mutation and returns the data.

This example demonstrates how we execute a single mutation; we will use this approach throughout the project to execute all the mutations.

Implementing queries with a Photo example

Next, we are going to learn how to implement a query operation, and we will look at how to handle GraphQL queries in Vue 3. To do this, follow these steps:

1. First, let's define the GET_PINS query inside the /graphql/photos/queries.js file to retrieve all the pins for a particular board:

    ```
    export const GET_PINS = gql`
     mutation getPins($size: Int, $skip: Int, $filters:
       PinFiltersInput) {
       getPins(size: $size, skip: $skip,
               filters: $filters) {
         id
         title
         url
       }
      }
     `;
    ```

2. Next, we will use the useQuery hook to execute this GraphQL query and return the data to a variable. As usual, we imported the GET_PINS query and the useQuery hook from their respective locations:

    ```
    <script setup>
    import { useQuery } from '@vue/apollo-composable';
    import { GET_PINS } from '../../graphql/photos/queries';
    ```

3. Next, we created a user-facing function called getBoardPins, which executes our query to retrieve and return the respective queries:

    ```
    const getBoardPins = () => {
     return getPins();
    };
    ```

4. Lastly, the `getPins` function executes the GraphQL query with the required variables and returns the result, as shown in the following snippet:

```
const { query: getPins } = useQuery(GET_PINS, () => ({
  variables: {
    size: 20,
    skip: 0,
    filters: {
      boardId: board.id,
    },
  },
}));
```

This is a typical example of how we can implement GraphQL queries throughout the code base for our Pinterest demo application.

If you followed along from each chapter, you will have cloned the repository from the URL provided in the previous section, set it up locally, and seeded some data into the Strapi backend server.

You should be presented with a Pinterest-like demo application, as shown here:

Figure 6.13 – Final Pinterest preview

> **Important note**
>
> The images might differ based on the data you seed into your Strapi backend database. However, the repository contains instructions on getting the dummy data and seeding them into your Strapi backend.

You can check out the complete implementation in the official GitHub repository here: `https://github.com/PacktPublishing/Architecting-Vue.js-3-Enterprise-Ready-Web-Applications/tree/chapter-6`. Additionally, you can learn how to implement the same API pattern with a RESTful API using the repository pattern here: `https://medium.com/backenders-club/consuming-apis-using-the-repository-pattern-in-vue-js-e64671b27b09`.

In conclusion, we created the frontend of our Pinterest clone using Vue 3 and the Composition API; we also created the backend using a very popular headless CMS called Strapi to store our data. Lastly, we integrated this into a single enterprise application using GraphQL.

Summary

This chapter dived deeper into how to utilize GraphQL to deliver scalable and high-performing applications and how to build a complete Pinterest clone with Vue 3 and GraphQL. In addition, we utilized the knowledge of GraphQL to develop and deliver enterprise applications such as Pinterest using Vue 3 and GraphQL.

We explored Strapi – the headless CMS that manages our backend APIs and data, and we also scaffolded a new Strapi project, learned how to create Strapi collections, and also seeded some dummy data to ease the development time.

We also covered in detail how to integrate the Strapi CMS and easily spin up a backend server for our Pinterest demo application using GraphQL into Vue 3.

In the next chapter, you will learn about the nitty-gritty involved in dockerizing your Vue 3 project. In addition, you will learn about the best practices and industry standards for dockerizing and deploying an enterprise Vue.js 3 web application.

This chapter will also go more practical by dockerizing a full-stack web application and deploying the container to a cloud platform using Docker Compose. Finally, you will learn how to handle larger projects with Docker Compose.

7

Dockerizing a Vue 3 App

In the previous chapter, we demonstrated how to build a complete Pinterest clone with Vue.js 3, GraphQL, and Strapi for the backend. You also utilized your knowledge of GraphQL to develop an enterprise Pinterest clone application. In this chapter, you will learn the nitty-gritty details of the steps involved in dockerizing your Vue.js 3 project. In addition, you will learn about best practices and industry standards to dockerize and deploy an enterprise Vue.js 3 web application. This chapter will also take a more practical approach by covering how to dockerize a full stack web application and deploy the container to a cloud platform using Docker Compose.

We will cover the following key topics in this chapter:

- Overview of Docker
- Dockerizing the app
- Running the app on Docker
- Dockerizing Vue.js 3 and Node.js with Docker Compose
- Running the app on Docker Compose

By the end of this chapter, you will have learned about best practices and industry standards to dockerize and deploy an enterprise Vue.js 3 web application. You will also have gained practical experience by dockerizing a full stack web application and deploying the container to a cloud platform using Docker Compose.

Technical requirements

To get started with this chapter, I recommend you read through *Chapter 6, Building a Complete Pinterest Clone with GraphQL*, first, where we built a complete Pinterest clone using Vue.js 3, GraphQL, and the Strapi CRM for the backend. We will be using that application a lot in this chapter to learn about Docker and Docker Compose.

All the code files for this chapter can be found at `https://github.com/PacktPublishing/Architecting-Vue.js-3-Enterprise-Ready-Web-Applications/tree/chapter-7`.

Overview of Docker

Docker has evolved over the years and knowing how to use it has become one of the most critical and in-demand skills for anyone interested in DevOps. Therefore, whether you're a seasoned DevOps engineer or a beginner, you definitely need to add Docker to your collection of skills.

Docker is the new buzzword in the DevOps and container orchestration industry. It was created in 2013 and was developed by the parent company, Docker, Inc.

Docker can package an application and its dependencies in a virtual container that can run on any Linux, Windows, or macOS computer. A container refers to an isolated or bundled application with the tools, libraries, and configuration files needed to execute the application.

One of the benefits of Docker is that it is a toolkit that enables developers to build, deploy, run, update, and stop containers using simple commands and work-saving automation through a single API across different operating systems and platforms.

This chapter explores everything you need to know about Docker, the different unique features of Docker, and why you should consider dockerizing your applications. We will also work through creating and setting up your first Docker application.

In the next section, we will explore Docker and its benefits to give us insights into why we need it in our development pipeline.

What is Docker?

Docker is an open source platform that allows developers to build, test, and deploy applications quickly. Docker achieves this by packaging your application in standardized units called containers. These containers have everything the software needs to run, including libraries, system tools, code, and a runtime environment. It also virtualizes the operating system of the computer on which it is installed and running.

To further explain this, let's say we have developed two different instances of our application, that is, the frontend and the backend.

The backend is developed with a Node.js stack, including a PostgreSQL database and other tools that make the Node.js backend execute properly on your local server.

Next, your frontend is created with Vue.js 3 and the necessary tools and configuration that make your Vue.js 3 application run smoothly.

Here are a couple of problems that might arise when working in a team or individually if you aren't using Docker:

- If a new team member joins, it might be tedious to onboard the member into the code base since the member needs to install and configure the correct version of the project and download the exact versions of the files required.

- When deploying the application, provisioning different servers for all the services used by your application will be a lot of work. For instance, you will have to provision different servers for the database, frontend, and backend. You may also need to provision different servers for different environments, such as staging, testing, and production, or use one server with lots of configurations each time.

With Docker, you can solve these problems by configuring, provisioning, and packaging all these services with a simple configuration file called a Dockerfile or a YAML file to define and run multi-container Docker applications using Docker Compose.

To understand how Docker will solve these problems, take a look at the following screenshot:

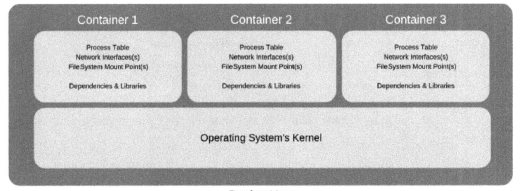

Figure 7.1 – A screenshot of the Docker host and layers (source: freeCodeCamp
[freecodecamp.org/news/docker-simplified-96639a35ff36/])

The preceding screenshot shows the internal layers and structure of Docker, where your application is bundled into containers with all the required resources to run smoothly. Additionally, each container uses your shared system resources.

This allows each container to be isolated from the others present on the same host. Thus, it allows multiple containers with different application requirements and dependencies to run on the same host, as long as they have the same operating system requirements.

Therefore, with Docker, you can run multiple applications as containers and use commands and a single configuration file to control everything. In the next section, we will carefully examine the benefits of Docker to our development pipeline.

The benefits of Docker

Some of the key benefits of using Docker are listed here:

- **Optimized storage system**: Containers are usually a few megabytes in size and consume very little disk space. Therefore, a large number of applications can be hosted on the same host.

- **Cost-effective**: Docker is less demanding when it comes to the hardware required to run it. Therefore, it reduces the cost of acquiring expensive hardware for different setups drastically.

- **Robustness**: Docker has a faster boot time as it consumes very little memory in comparison to a virtual machine since it does not have an operating system installed.

- **Multiple containers**: With the same operating requirements, Docker supports multiple applications with different application requirements and dependencies, to be hosted together on the same host.

These are the benefits of using Docker to manage and ship your enterprise applications. Let's next explore why you should use Docker in your enterprise-level application.

Why use Docker

Docker enables you to publish your code quickly and efficiently. It standardizes the operations of an application, allows you to move code seamlessly, and saves revenue by improving the utilization of resources.

Here are some of the reasons you should start using Docker in scalable enterprise applications:

- **Ship more software faster**: According to Amazon (`https://aws.amazon.com/docker/`), Docker users ship products 7x faster than non-Docker users. Docker enables you to ship isolated services as often as needed. When building enterprise-level and scalable applications, features and bug fixes happen in hours, if not minutes. Therefore, urgent building, testing, and deployment are needed and Docker comes in handy in this area.

- **Standardize operations**: Docker follows industry-standard application development practices. Isolated standardized units called containers make it easy to deploy, identify issues, and roll back for remediation.

- **Seamlessly move**: Developers can move applications between different environments and systems without worrying about installing any libraries or missing configuration files. Docker-based applications can be moved seamlessly from local development to a production environment.

- **Save money**: Docker-based applications are cost-effective since you can run multiple applications on one server in the form of containers. Docker containers make it easier to run more code on each server, improving your utilization of CPU resources and saving you money.

Now you know why you should use Docker in your enterprise applications and the benefits you can incur from using Docker. With Docker, you can ship products faster and more efficiently. When combined with other industry-standard tools, you can completely remove the hassle of manual deployment by instead adopting automated deployment. In the next section, we are going to explore how to dockerize your first application.

Implementing Docker with Vue.js 3

Docker is an enterprise-ready container platform that enables organizations to seamlessly build, share, and run any application, anywhere. Almost all enterprise-level companies containerize their applications for faster production workloads so that they can deploy anytime, sometimes several times a day.

One way to build an enterprise-level application is to dockerize the project from the beginning. Therefore, we are going to dockerize the Pinterest Vue.js 3 app with the Strapi backend we developed in *Chapter 6, Building a Complete Pinterest Clone with GraphQL*, and create a Docker image so that we can deploy that image any time or sometimes several times a day.

Prerequisite

Most importantly, you must download and install Docker in your local development system for local testing. You can go to this link to download and install it on different operating systems: `https://docs.docker.com/install/`.

Example project

In *Chapter 6, Building a Complete Pinterest Clone with GraphQL*, we developed a Pinterest clone using Vue.js 3 and Strapi for the backend. In this section, we will learn how to dockerize the project from scratch. Here is a demo of the application:

Figure 7.2 – A screenshot of the Pinterest clone demo

The application displays images in a masonry grid layout based on the number of images we have stored in our Strapi database.

The Strapi backend allows you to manage and control the entire backend of the application, from adding pins and boards to creating new users.

We will dockerize both the Vue.js 3 Pinterest app and the Strapi backend using individual Dockerfiles and multi-stage builds to create efficient Docker images.

Dockerizing the Pinterest app

We will start by dockerizing the Pinterest Vue.js 3 application. In this multi-stage build, building a Vue.js 3 project and putting those static assets in the dist folder is the first step. So, let's create a Dockerfile and configure our Vue.js 3 application.

Create a Dockerfile in the root directory of your Pinterest clone Vue.js 3 app. If the Strapi backend is still inside the frontend folder as you clone from the `Chapter 6` repository (`https://github.com/PacktPublishing/Architecting-Vue.js-3-Enterprise-Ready-Web-Applications/tree/chapter-6`), you can create a parent folder and move the Strapi backend folder side by side with the frontend folder, as shown in the following screenshot:

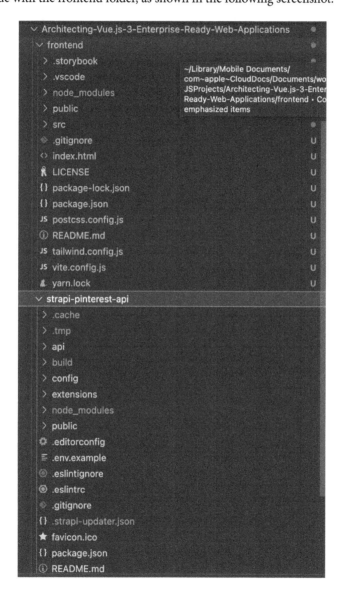

Figure 7.3 – A screenshot of the current folder structure

Moreover, you can clone the completed `Chapter 7` (`https://github.com/PacktPublishing/Architecting-Vue.js-3-Enterprise-Ready-Web-Applications/tree/chapter-7`) repository from this link, which contains the complete code base.

Lastly, let's explore the configuration file to dockerize the Vue.js 3 enterprise application. Open your Dockerfile and add the following code:

```
# Use the official Node.js 14 Alpine image from https://hub.
docker.com/_/node.
# Using an image with specific version tags allows
deterministic builds.
FROM node:16.17.0-alpine

# Create and change to the app directory.
WORKDIR /usr/src/frontend

# Copy important root files to the builder image.
COPY package*.json ./

# Install production dependencies.
RUN npm install

# Copy the Vue 3 source to the container image.
COPY . .

# Expose container port
EXPOSE 3000

# Run the Vue service on container startup.
CMD ["npm", "run", "dev"]
```

The code snippet is self-explanatory with the comments explaining every command we used in the Dockerfile.

Let's build the image with the following command:

```
// build the image
docker build -t pinterest-vue-frontend .
// check the images
```

```
docker images
```

In summary, we have successfully dockerized our Vue.js 3 Pinterest application. In the next section, we will dockerize the Strapi backend application separately.

Dockerizing the Strapi backend app

In this section, we will follow the same approach used in dockerizing the Vue.js 3 frontend project to create a Docker instance for the Strapi backend.

Therefore, create a Dockerfile inside the Strapi backend folder of your project and add the following configuration code:

```
# Use the official Node.js 14 Alpine image from https://hub.
docker.com/_/node.
# Using an image with specific version tags allows
deterministic builds.
FROM node:14.16.1 AS builder
# Create and change to the app directory.
WORKDIR /usr/src/backend
# Copy important root files to the builder image.
COPY package*.json ./
# Install production dependencies.
RUN npm install
# Copy the Backend source to the container image.
COPY . .
# build app for production with minification
RUN npm run build
EXPOSE 1337
# Init final image generation.
FROM node:14.16.1
# Run the Strapi service on container startup.
CMD ["npm", "start"]
```

We copied the previous configuration and changed the building process for the backend app. The code snippet is self-explanatory with comments explaining every command we used in the Dockerfile. In the next section, we are going to run the applications on Docker and test them separately.

Running the images on Docker

After building the Docker image, next, we need to run the image on Docker using the following command:

```bash
```bash
// run the Frontend image
docker run -d -p  3000:3001 --name pinterest-frontend
pinterest-vue-frontend
// run the Strapi Backend Image
docker run -d -p  1337:3002 --name pinterest-backend pinterest-
strapi-backend
// check the container
docker ps
```
```

The ps command checks the container for the list of images currently running in your Docker engine. You should see two images with the names specified in the preceding Docker run command.

If the run command is successful, you can access the frontend application on the web at the address http://localhost:3001 and the backend instance at port 3002. This port change is possible because the -p option exposes our internal frontend Vue.js 3 server port 3000 to the external port 3001, which makes it possible to access our internal Docker application in our browser.

At this point, if everything is successful, you should be greeted with your demo Vue.js 3 application. However, following this approach poses a problem. Developers need to build, test, and deploy applications in isolation, which can be avoided with Docker Compose.

In this section, we explored how to dockerize the Pinterest clone application we have developed in this book. We learned how to create, build, and run the Dockerfile we used in dockerizing the project using different Docker commands. In the next section, we will explore how to use Docker Compose to build, test, and deploy multiple applications at once.

Dockerizing Vue.js and Node.js with Docker Compose

In the previous section, we explored how to dockerize Vue.js 3 applications and how to dockerize a Node.js application using Strapi, which was done separately. In this section, we are going to explore how to build, test, and deploy bundled applications. Furthermore, we are going to build and dockerize both applications as a single unit.

Overview of Docker Compose

Docker Compose is a tool designed to enable users to easily define and share multi-container applications. By creating a YAML file, Compose allows us to quickly launch or shut down all services with a single command.

With Docker Compose, developers can build, test, and deploy multiple containers and images bundled together to form a single application.

In the next section, we are going to explore how to bundle the frontend and backend applications that we demonstrated in the previous section.

Dockerizing the Pinterest clone app

To bundle a deployable application, we are going to start by creating a central Dockerfile and Docker Compose YAML file inside the root directory that contains the different configurations to bundle our application with Docker Compose.

Before you start, rename your `strapi-pinterest-api` folder to `backend`. The following screenshot shows the current folder structure:

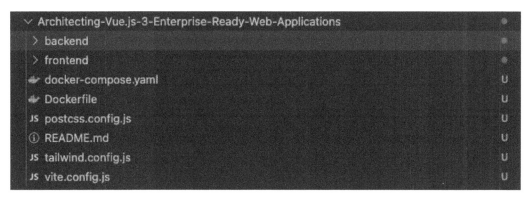

Figure 7.4 – A screenshot of the current folder structure with Docker and Docker Compose files

Next, create a Dockerfile inside the root directory and add the following script:

```
FROM node:14.15.0
ARG PACKAGE_PATH=
ARG WORKING_DIR=
WORKDIR ${WORKING_DIR}
COPY ${PACKAGE_PATH}/package*.json ${WORKING_DIR}
RUN npm install --silent
COPY ${PACKAGE_PATH} ${WORKING_DIR}
```

```
VOLUME $WORKING_DIR/node_modules
CMD [ "npm", "start" ]
```

Code walk-through

Let's walk through the code together and understand the nitty-gritty of it.

Step 1: Import Node.js

The first step in every Dockerfile is to specify the build image. In this case, we specify Node.js as our image.

This will install Node.js with the specified version number and set up the environment to run Node.js properly.

Step 2: Create the required arguments

The second step is to create the arguments required by Docker Compose when building individual images of our frontend and backend applications:

```
ARG PACKAGE_PATH=
ARG WORKING_DIR=
WORKDIR ${WORKING_DIR}
```

Additionally, we create a working directory specifying the argument we created earlier. This will auto-inject the specified working directory in the Docker Compose YAML file.

Step 3: Copy, install, and run commands

Lastly, we copy files from the specified working directory into the Docker virtual working directory. We start by copying package*.json files, running the install command, and copying the remaining files later. This approach utilizes the Docker caching system:

```
COPY ${PACKAGE_PATH}/package*.json ${WORKING_DIR}
RUN npm install --silent
COPY ${PACKAGE_PATH} ${WORKING_DIR}
VOLUME $WORKING_DIR/node_modules
CMD [ "npm", "start" ]
```

Furthermore, after mounting the node_modules folder of the specified working directory, Docker will execute the npm start command to start the application.

Now that we have created a central Dockerfile for both the frontend and backend, let's continue by creating a YAML Docker Compose file to bundle our separate applications together.

Create a `docker-compose.yaml` file inside the root directory and add the following script:

```yaml
version: "3.5"
services:
 api:
   build:
     context: .
     dockerfile: Dockerfile
     args:
       PACKAGE_PATH: backend
       WORKING_DIR: /usr/src/
   expose:
     - 1337
   ports:
     - 1337:1337
   environment:
     - NODE_ENV=development
     - HOST=0.0.0.0
     - PORT=1337
     - BASE_URL=http://api:1337
   env_file:
     - ./.env
   volumes:
     - ./backend:/usr/src
   command: >
     sh -c "npm install"
 frontend:
   build:
     context: .
     dockerfile: Dockerfile
     args:
       PACKAGE_PATH: frontend
       WORKING_DIR: /usr/src/
   expose:
     - 3000
   ports:
     - 3000:3000
```

```
environment:
  - APP_ENV=production
  - APP_BACKEND=http://0.0.0.0:1337/api
  - NODE_PATH=/usr/src/
  - APP_TOKEN=eyJhbGciOiJIUzI1NiJ9.c29sb[STRAPI_TOKEN]
env_file:
  - ./common.env
volumes:
  - ./frontend:/usr/src
depends_on:
  - api
command: ["npm", "start"]
```

Code walk-through

Let's walk through the code together and understand the nitty-gritty of it.

Step 1: Versioning and services

Every Docker Compose file always starts with a version number of the version of Docker Compose you intend to use when building and bundling the application. In this demo, we specify version 3.5.

Furthermore, every Docker Compose YAML is always split into different services. You can add as many services as required that each application depends on. For instance, if the backend of your project depends on a database (PostgreSQL), you can specify that as a service.

In this demo, we have specified only two services, namely the following:

- Backend
- Frontend

Each of the services contains configurations that enable them to run smoothly. Let's explore the configurations we have added to the frontend service.

Step 2: The build section

The build section includes configurations that help in building the application. It contains commands such as the context, working directory, and defined arguments:

```
build:
  context: .
  dockerfile: Dockerfile
  args:
```

```
    PACKAGE_PATH: frontend
    WORKING_DIR: /usr/src/
```

The `context` command specifies the part of the directory where the Dockerfile we created earlier is stored. In our case, it was stored in the root directory.

Next, we call the Dockerfile with the `dockerfile` command and specify the required parameters with the `args` command. Lastly, we specify the `PACKAGE_PATH` and `WORKING_DIR` values.

Step 3: Exposing the port

In this step, we exposed the internal Docker port used to run the application to the outside world:

```
expose:
  - 3000
ports:
  - 3000:3000
```

Step 4: Creating environment variables

In this step, we use the `environment` command to add the required environment variables, such as `APP_BACKEND` and `APP_ENV`:

```
environment:
    - APP_ENV=production
    - APP_BACKEND=http://0.0.0.0:1337/api
    - NODE_PATH=/usr/src/
    - APP_TOKEN=eyJhbGciOiJIUzI1NiJ9.c29sb[STRAPI_TOKEN]
  env_file:
    - ./.env
```

Then, we create an env file in the root directory using the `env_file` command to store the details we specified previously.

Step 5: Running the app

Lastly, we mount the `frontend` directory and specify that the frontend project depends on our backend API service, which is our Strapi backend. This allows Docker to execute the project in sequence from the backend first before the frontend:

```
volumes:
    - ./frontend:/usr/src
  depends_on:
```

```
        - api
    command: ["npm", "start"]
```

Finally, we call the command to execute the project. This same approach is repeated for the backend service. In the next section, we are going to learn how to run the project using Docker Compose.

Running the app on Docker Compose

After creating a successful Docker Compose YAML configuration file, let's run our Pinterest clone project using Docker Compose.

Before you start running the project, make sure to set up and install Docker and Docker Compose. Next, type `docker compose up` in your terminal root directory to deploy the project. Alternatively, type `docker-compose up` to use Docker Compose directly. The application will be served at `http://localhost:3000/`.

If everything is properly configured, you should be presented with a full stack Pinterest application, as shown in the following screenshot:

Figure 7.5 – Preview of Pinterest application demo

If you are unsure about anything, please refer back to the code base of this chapter (`https://github.com/PacktPublishing/Architecting-Vue.js-3-Enterprise-Ready-Web-Applications/tree/chapter-7`) to see the working and complete project setup.

In this section, we explored how to use Docker Compose to create and manage many services using the Dockerfile that we created in the previous section. We also learned how to bundle our full stack application, including the frontend, backend, and a database, using Docker Compose.

Summary

This chapter dove deeper into the nitty-gritty steps involved in dockerizing your Vue.js 3 project. In addition, we explored best practices and industry standards to dockerize and deploy an enterprise Vue.js 3 web application. We also learned how to dockerize a full stack web application using Docker Compose.

Using a Dockerfile, we were able to dockerize our Pinterest clone demo application so that it can be deployed and managed by other team members or on any cloud provider easily. Also, we learned how to bundle and manage a full stack application that includes the backend, the frontend, a database service, as well as many more features, all in a single file, using Docker Compose.

In the next chapter, you will explore the concept of testing. You will learn what to test from an array of available components and methods. In addition, you will learn about best practices and industry standards related to testing libraries and how to integrate them with Vue.js 3.

Part 4: Testing Enterprise Vue.js 3 Apps

Testing an enterprise project can be daunting and unnecessarily complex. This part will explore everything relating to enterprise testing and what to test precisely to eliminate time spent on testing the wrong code.

This part comprises the following chapters:

Testing and What to Test in Vue.js 3

In the previous chapter, you learned the nitty-gritty details of the steps involved in dockerizing your Vue.js 3 project. In addition, you learned about the best practices and industry standards to dockerize an enterprise Vue.js 3 web application.

In this chapter, you will explore the concept of software testing. You will learn what to test from an array of available components and methods. In addition, you will learn about best practices and industry standards related to testing libraries and how to integrate them with Vue.js 3.

We will cover the following key topics in this chapter:

- Overview of testing
- Testing in software engineering
- What to test
- Testing a basic Vue.js 3 app
- Component testing in Vue.js 3

Technical requirements

To get started with this chapter, I recommend you read through *Chapter 7, Dockerizing a Vue 3 App*, first, where we took a more practical approach by dockerizing a full stack web application using Docker Compose. We will be using the application a lot in this chapter to learn about Vue.js 3 enterprise testing.

All the code files for this chapter can be found at `https://github.com/PacktPublishing/Architecting-Vue.js-3-Enterprise-Ready-Web-Applications/tree/chapter-8`.

Overview of testing

Anyone who has studied computer science should be familiar with the concept of SDLC. If you are not aware, **SDLC** stands for **software development life cycle**.

Synopsys (`https://www.synopsys.com/glossary/what-is-sdlc.html`) provides the following definition:

> *Software Development Life Cycle (SDLC) is a structured process that enables the production of high-quality, low-cost software, in the shortest possible production time. The goal of the SDLC is to produce superior software that meets and exceeds all customer expectations and demands.*

If you explore the SDLC further, you'll see that it defines and outlines eight detailed plans with stages or phases that quality and enterprise-level software must pass through to produce software that meets and exceeds all customer expectations and demands.

Each stage is crucial, including planning, coding, building, and testing. However, the testing phase is particularly important, especially when you need to build a bug- or defect-free enterprise-level application.

To elaborate further, the testing phase evaluates the created software against any bugs, any potential errors, and the requirements of the software from the planning phase.

Next, we will see what we mean by software testing.

What is software testing?

Software testing is the method of checking whether the software in production matches the expected requirements and, most importantly, whether it is defect free. The method used to carry out software testing differs depending on the organization. However, the method is divided into manual and automated processes.

Individuals and organizations will have different names for software testing. It can also be categorized as whitebox or blackbox testing. However, the end result of any approach is always the same, which is identifying errors, gaps, or missing requirements in contrast to actual requirements.

Blackbox testing involves testing a system without knowing the internal workings of the system, while whitebox testing is an approach of testing that allows the tester to inspect and verify the internal workings of the system.

Regardless of the names, terms, or categories used when referring to software testing, in simple terms, software testing means the verification of the **Application Under Test** (**AUT**), and it's a critical and crucial stage in producing high-quality enterprise-level software.

In the next section, we will go through the importance of software testing.

Why software testing is important

The need to incorporate software testing into your application development pipeline cannot be overemphasized. It is as important as the planning and development phases in the SDLC. In fact, without a proper software testing strategy, it is likely that the end product of the software under development will be filled with bugs, errors, and unmet software requirements.

Software testing is important because software bugs could be expensive and also very dangerous to businesses, and organizations at large. Over the years, there have been numerous examples of potential software bugs and monetary losses.

For instance, in April 2015, the Bloomberg terminal in London crashed due to a software glitch that affected more than 300,000 traders in financial markets. It forced the UK government to postpone a 3 billion pound debt sale, according to The Guardian (`https://www.theguardian.com/business/2015/apr/17/uk-halts-bond-sale-bloomberg-terminals-crash-worldwide`).

Also, according to Windows Report (`https://windowsreport.com/windows-10-vulnerability/`), there was a vulnerability in Windows 10 that enabled users to escape from security sandboxes through a flaw in the win32k system.

There have been many vulnerability reports on different malicious attempts on businesses that have impacted revenue or monetary value, of which some could be avoided with proper software testing. This indicates that software testing is a very important stage in the SDLC.

Though software testing costs money, I'm sure you will agree with me that the cost is nothing compared to the millions per year in development and support companies would have to spend if they don't have a good testing technique and QA processes in place.

In addition, having early software testing in place uncovers problems before the products go to market. Early testing also uncovers different defects, including but not limited to the following:

- Architectural flaws
- Poor design decisions
- Invalid or incorrect functionality
- Security vulnerabilities
- Scalability issues

Having proper testing processes in place along the software development pipeline improves software reliability and means high-quality applications are delivered with few errors. In the next section, we will further explore the benefits of software testing.

The benefits of software testing

In the previous section, we explained why enterprise applications need to include testing in their development pipeline. In this section, we will explore the benefits of having a proper software testing process. We will go through the following points in detail:

- Helps in saving money
- Satisfaction of customers
- Enhancing the development pipeline
- Quality of product
- Security

Helps in saving money

Launching buggy software to market can end up being more expensive than creating the entire software. As stated previously, there have been many cases of company's monetary value reducing due to software defects and error-prone software.

This problem can be curtailed to some extent if there is a proper software testing process built into the development pipeline to detect and rectify these errors before moving on to the next stage of the pipeline.

Satisfaction of customers

For users of your enterprise application to be satisfied, the software must work properly and in accordance with the requirements.

Therefore, before launching the software to market, an acceptance test must be conducted to ensure that the product works in accordance with the requirements and also to get a sense of how users will access and use the product day to day.

While in the testing phase, if any issues or bugs are detected, the software under test can easily be moved back to the development stage instead of finding out about the problems in the production stage where real users are interacting with the application.

Enhancing the development pipeline

Including software testing in the development pipeline creates an enhanced development pipeline. Also, it is an industry-standard practice to include software testing. In addition, it is simpler for developers to fix errors in the development stage than in the production stage.

Thus, incorporating the software testing process in the development pipeline reduces the risk of launching error-prone software to market and enhances the development pipeline.

Quality of product

When the quality of the product drops, the company might lose customers, resulting in a loss of revenue. However, one of the attributes of low-quality software is an untested and error-prone software application.

Furthermore, if there is a proper software testing process built into the development pipeline, most errors, bugs, and defects will be detected and fixed before production, thereby producing quality software.

Security

According to OWASP (`https://owasp.org/www-project-top-ten/`), security should be an integral part of every software. Not considering it could result in a reduction in the monetary value of the business.

Software testing is one way to detect security loopholes and fix them in the development stage. If a product has undergone testing, the user can be assured that they are receiving a reliable product. They will be assured that their personal details are safe. Users can receive products that are more likely to be free from vulnerabilities with the aid of software testing.

In this subsection, we covered some benefits you can derive from implementing software testing into your pipeline when building enterprise-ready applications.

In the next section, we will explore the different types of testing and different strategies you can integrate into your development pipeline.

Testing in software engineering

As stated in the previous section, software testing is an integral part of the SDLC, and therefore, according to ANSI/IEEE 1059, testing in software engineering is a method of evaluating the software under test to discover whether it meets the requirements, as well as whether it is error, bug, and defect free.

The process involves evaluating the features of the software under test for requirements in terms of any missing requirements, bugs or errors, security, reliability, and performance.

In this section, we explored the benefits of testing and why software testing is important, and in the next section, we will understand different types of software testing. We will explore what to test and how to write basic unit and integration tests.

Types of software testing

Software testing has been given different names. There are over 150 types of software testing according to Guru99 (`https://www.guru99.com/types-of-software-testing.html`).

However, we are going to classify software testing into two main categories and then explore each of the categories and the different types within them. The following are the two main categories:

- Functional software testing

- Performance software testing (non-functional)

The following figure shows the high-level classification of software testing types:

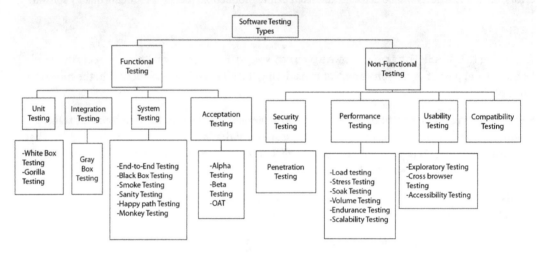

Figure 8.1 – A high-level classification of software testing (source: softwaretestinghelp)

You can explore more about the different categories of software testing on Youtube. However, we will only focus on three main categories of functional software testing, as follows:

- Unit testing

- Integration testing

- End-to-end testing

Unit testing

This basic approach to software testing is carried out by a programmer to test the unit or smallest part of the program. It helps developers to know whether individual units of code are working properly or not.

Integration testing

This type of testing focuses on the construction and design of the software. You need to see whether the integrated units are working without errors or not.

End-to-end testing

End-to-end testing is a methodology that assesses the working order of a complex product in a start-to-finish process.

In the coming chapters, we will focus on exploring these different types of testing individually. Nevertheless, you can explore more than 150 different types of software testing from Guru99 (`https://www.guru99.com/types-of-software-testing.html`).

In summary, now that we know how important software testing is and the different types of software testing, how do we know what to test in a large enterprise application? In the next section, we are going to explore what to test and how to integrate a testing pipeline into the development workflow.

What to test

A popular question among software teams is *what should we test and what should we not test?* In this section, we will explore the different things you should and shouldn't test when considering software tests.

We will first explore different test strategies to employ when integrating software testing into your development workflow.

Testing strategy

The best testing strategy to implement in your enterprise application is the combination of normal (manual) testing and automated testing. In addition, normal testing should be done more extensively by the **Quality Assurance (QA)** team.

To explain this further, when automated testing is written and implemented successfully, we usually program it to look for fundamental errors and edge cases that may not properly assimilate how a real customer will interact with the application.

What you should test

As much as software testing is important to the efficiency of an enterprise-ready application, knowing what to test is paramount so that developers don't waste time testing the wrong things.

The following is a list of some of the things you can look for when testing your enterprise project for errors, bugs, and defects:

- **Passed parameters**: The collection of parameters or arguments passed into the method or function to make sure that it has not changed. In some cases, the data type of the parameter remains the same.

- **Algorithm engines**: Every method has a purpose, and the purpose is implemented using logic or an algorithm. Your test case should test the algorithm to make sure it's correct and it results in the right output based on the input into the method.

- **Simple database queries checking predicates**: If your job as a developer is related to queries and manipulating databases, you really want to test your database queries to make sure it performs the right manipulation and queries.

- **Utility methods**: Utility methods are helpers in your project that are created for a specific task. They are usually used when you need to do stuff that does not need an instance of a class. This set of methods needs to be tested properly to ensure it produces the correct output when used.

- **Testing less critical codes:** Test the edge cases of a few unusually complex pieces of code that you think will probably have errors. Additionally, carry out edge-case tests of less critical code whenever someone has time to kill.

The preceding are a few things you can consider for your test cases. However, it is important to note that writing tests and having 100% code coverage do not necessarily mean that your code is bug free. In the next section, we will explore things you should not test in your project.

What you should not test

The following are the things that you should not be testing in your project:

- Constructors or properties (if they just return variables). Test them only if they contain validations.

- Methods that call another public method.

- If the code needs to interact with other deployed systems, then an integration test should be used.

- Configurations such as constants, read-only fields, configs, and enumerations.

- You should not test POJO classes or models; rather, you can test each of the methods inside the class.

In summary, we have explored the software testing strategy, what to test, and what you should not test to help you understand the relevance of software testing in your enterprise application. In the next section, we will explore how to test a basic Vue.js application.

Testing a basic Vue.js 3 app

In the previous chapter, we created a Pinterest application using Strapi for the backend and Vue.js 3 for the frontend.

Previously, we added internationalization, structured the project, and built out a complete Pinterest clone. In this section, we will continue by using the official project we created for this book to set up software testing, resulting in a full-blown enterprise-ready Pinterest clone application.

You can clone the project from this repository, `https://github.com/PacktPublishing/Architecting-Vue.js-3-Enterprise-Ready-Web-Applications`, to jump right in.

Creating a test folder

It is always confusing knowing where to add your test files and folder when it comes to creating enterprise applications. There are two methods for structuring your test files depending on the approach you used for your enterprise project.

Method 1 – adding test files inside each component

First, you can create a specific test file inside each of the `component` folders. For instance, inside the `component` folder in our Vue.js 3 project, we will create a folder for each component and move the files of each component into the folder, including the test file for each component.

The following figure shows an example of how we could arrange our component folder to accommodate our testing files and other files related to a particular component:

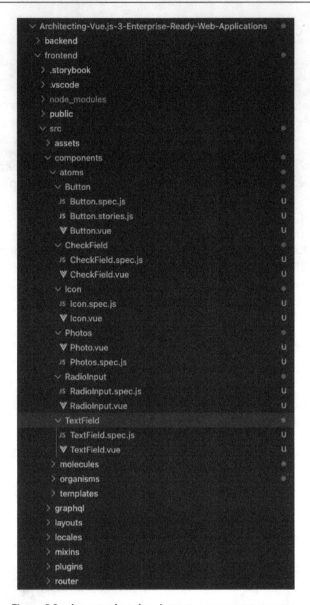

Figure 8.2 – A screenshot showing our component structure

In the preceding figure, you can see how you can add any files related to any of the components. For example, if you want to add an end-to-end testing or integration testing file for each component, you just create the file within each specific component folder.

Also, because of the structure of our practice project and the introduction of the atomic pattern, we can easily see how many files will be created in each component. The same goes for different areas we will be testing throughout the project.

However, we can use the next method to arrange everything related to testing in a separate folder and create all the files and folders inside the specific `tests` folder.

Method 2 – creating a tests folder

In this section, we will create a folder inside the `src` folder called `tests`, which will contain every file and folder related to testing and test configurations.

The following figure shows the folder structure for implementing testing with this method:

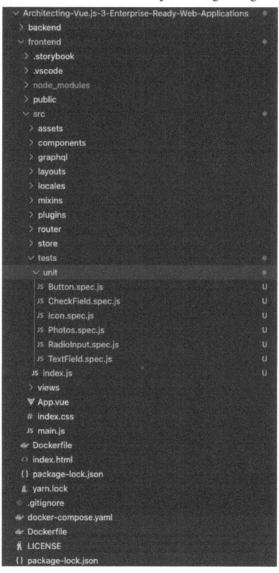

Figure 8.3 – A screenshot of the folder structure

This method has a single folder that contains all the files and folders related to testing, including all the configurations. It is a central place and single point of truth for all your software testing files and folders. You can arrange this folder in a different structure as per your use case.

Furthermore, the method of structuring your test folder or files and folders related to testing does not matter. What matters the most is implementing software testing properly and knowing exactly what to test to avoid production bugs and errors.

We will use the second method in this demo in writing some basic unit tests to demonstrate. This is because we don't want the testing files to be scattered across different folders since we are using the atomic pattern.

Writing a basic unit test

First, we will start by installing the new testing library for Vue.js 3. Since we are using Vite in the project, we will also install the Vitest library for our test runner. You can read more about the new Vue.js 3 test library at `https://vitest.dev/guide/`.

Installing the test library

Vitest is a blazing-fast unit test framework powered by Vite. Install the library by using any of these commands:

```bash
# with npm
npm install -D vitest

# or with yarn
yarn add -D vitest

# or with pnpm
pnpm add -D vitest
```

Now that we have our testing library set up, let's create a simple helper file to test our configuration.

Creating a helper file

For our demonstration, we will create a helper file inside the `src/helpers` folder and add a simple function to increment a value. The following snippet shows the code example we have added to the newly created file:

```
// src/helpers/index.js
export function increment(current, max = 10) {
  if (current < max) {
    return current + 1;
  }
  return current;
}
```

The increment function written previously increments a value by 1 until the max value is reached. If the max value is reached, it simply returns the current value. Next, let's write a basic unit test for it.

Writing a basic test

In this section, we will write a simple unit test for this function. You can follow this by writing unit tests for all the functions and methods of your enterprise project:

```
import { describe, it, expect } from 'vitest';
import { increment } from '../../helpers';

describe('increment', () => {
  it('increments the current number by 1', () => {
    expect(increment(0, 10)).toBe(1);
  });

  it('does not increment the current number over the max', () =>
{
    expect(increment(10, 10)).toBe(10);
  });

  it('has a default max of 10', () => {
    expect(increment(10)).toBe(10);
  });
});
```

We will explore unit testing in depth in the next chapter.

Let's walk through the preceding code together and understand the nitty-gritty details of it:

- **Step 1: Adding the required packages**

  First, we need the `vitest` package and the `helper` file to test. Next, we use the exported functions to create a `describe` block, and so on:

  ```
  import { describe, it, expect } from 'vitest';
  import { increment } from '../../helpers';
  describe('increment', () => {

  ….

  });
  ```

  The `describe` block is used to group related test cases, as demonstrated in the preceding code snippet.

- **Step 2: Using the it function**

  Next, we use the `it` function to test specific use cases of our function. For instance, we test to make sure the number actually increases by `1` each time the function is called:

  ```
  it('increments the current number by 1', () => {
      expect(increment(0, 10)).toBe(1);
  });
  ```

- **Step 3: Using the except function**

  Lastly, the `expect` function is used to test the use case. You pass in a value and expect the value to be equal to another value, as shown in the example.

You can access different methods (`https://vitest.dev/api/#expect`) from the `expect` object aside from the `toBe()` function.

In the next section, we will cover the process of running your test with Vitest and creating your first component/integration testing examples.

Running a test with Vitest

We will now run the test to see whether it passes or not. Type the following command into your root terminal:

```
yarn test
```

If your test is successful, you should see that the three cases passed, as shown in the following figure:

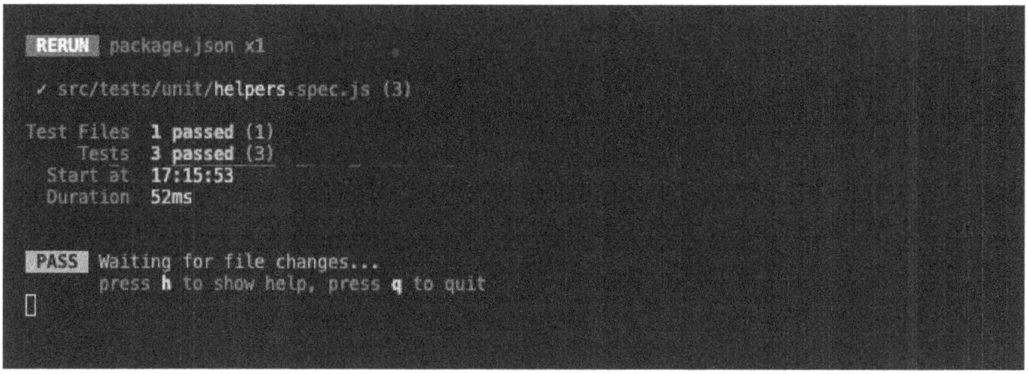

Figure 8.4 – A screenshot showing the test result

In this section, we have demonstrated how to configure and structure software testing with Vue.js 3 using the latest Vitest testing library for Vue.js 3. We have also learned how to write a basic unit test. In the next section, we will learn how to create basic component-based testing.

Component testing in Vue.js 3

In Vue.js, components are the main building block of the UI and refer to a single unit of the application that is shareable, testable, and reusable. Therefore, component testing sits between unit testing and end-to-end testing. It can also be referred to as integration testing.

In the previous chapters, where we integrated atomic patterns using Storybook, we discussed creating component stories and how to create them. If you created stories for all your components and configured them to work properly as demonstrated, then you will have already implemented component testing using Storybook stories.

However, *Chapter 10, Integrating Testing in Vue.js 3* is dedicated to exploring component testing. Nevertheless, we will briefly illustrate in this chapter how to implement simple component-based testing to aid our understanding of the next chapters.

Writing a basic component test

We will start by installing the new testing library for Vue.js 3. Since we are using Vite in the project, we will also install the Vitest library for our test runner. You can read more about the new Vue.js 3 test library (https://vuejs.org/guide/scaling-up/testing.html).

Installing the test library

As of the time of writing, `@testing-library/vue` (https://github.com/testing-library/vue-testing-library) is recommended for component testing, and we will install it in our example. Run the following command to install it:

```bash
npm install -D vitest happy-dom @testing-library/vue
```

Next, open the `vite.config.js` file and add the following configuration:

```
import { defineConfig } from 'vite';
import vue from '@vitejs/plugin-vue';
// https://vitejs.dev/config/
export default defineConfig({
 plugins: [vue()],

 test: {
   environment: 'happy-dom',
   globals: true,
 },
});
```

This configuration should allow us to run both unit and component tests at the same time. Next, let's create a simple component test.

Creating a basic component test

Here is a simple component test snippet to demonstrate the process:

```
import { render } from '@testing-library/vue';
import Button from '../../components/atoms/Button.vue';
test('mounted a button with custom label', async () => {
  // The render method returns a collection of utilities to
  // query your component.
  const { getByText } = render(Button, {
```

```
    props: {
      label: 'Test',
    },
  });
  // getByText returns the first matching node for the
  // provided text, and Check if button is render with Label
  // from props
  const button = getByText('Test');
});
```

The test simply renders Button with a custom label value and also checks whether we can retrieve the custom label added during the rendering process.

Running the test

Running the test will result in four passed test cases, including the unit test we created earlier:

Figure 8.5 – A screenshot showing the final test result with integration testing

In this section, we have demonstrated how to configure and structure component testing, also known as integration testing, with Vue.js 3 using the latest Vitest testing library for Vue.js 3. We have also written basic component tests to help us understand the process. In the next chapters, we will look in-depth at the different types of testing we can perform when building enterprise projects with Vue.js 3.

You can clone the latest repository for this chapter here: https://github.com/PacktPublishing/ Architecting-Vue.js-3-Enterprise-Ready-Web-Applications/tree/chapter-8.

Summary

This chapter dove deeper into the concept of software testing to deliver scalable, high-performing, and bug-free applications. We also explored what to test from an array of available components and methods. In addition, we utilized our knowledge of software testing to create basic unit and component test cases in Vue.js 3 using Vitest.

In the next chapter, we will explore everything related to unit testing. We will learn how to unit test a Vue.js 3 component and the component and page methods. We will also learn about unit testing tools such as Vitest and use them to effectively unit test an enterprise project.

9

Best Practices in Unit Testing

In the previous chapter, we learned about the concept of software testing. We learned what to test from an array of available components and methods. In addition, we learned about the best practices and industry-standard testing libraries and how to integrate them with Vue.js 3.

In this chapter, we will explore everything related to unit testing. We will learn how to unit test a Vue.js 3 component and the component and page methods. We will also learn about unit testing tools such as Jest and Mocha and how to use them to effectively unit test an enterprise project.

We will cover the following key topics in this chapter:

- Introduction to unit testing

- What is unit testing?

- The importance of unit testing

- The benefits of unit testing

- Best practices in unit testing

- JavaScript unit testing

- Testing a basic Vue.js app

Technical requirements

To get started with this chapter, I recommend you read through *Chapter 8, Testing and What to Test in Vue.js 3*, first, where we explored the concept of software testing and what to test from an array of available components and methods. We will rely heavily on the knowledge from that chapter when learning about Vue.js 3 enterprise unit testing in this chapter.

All the code files for this chapter can be found at `https://github.com/PacktPublishing/Architecting-Vue.js-3-Enterprise-Ready-Web-Applications/tree/chapter-9`.

Introduction to unit testing

Unit testing is an important step in the development phase because it helps spot errors and defects at the development stage if done correctly.

Unit testing is a method of software testing in which the smallest testable parts of the software under test, called units, are individually or independently tested for proper operation and to make sure the output corresponds with the required output.

The units can be said to be individual functions, objects, methods, procedures, or modules in the software under test.

This software testing approach is developed by software engineers to test the units of the program. It helps software engineers to know whether individual units of the code are working properly or not.

In this section, we will examine the definition of unit testing and explore how developers can create and manage unit tests within their enterprise-ready Vue.js 3 applications.

What is unit testing?

Unit testing is a method of verifying the smallest piece of testable code against its purpose or requirement. This method makes sure that the smallest part of your code base is tested and made to conform to the requirements.

It is very important to discover and fix bugs during the development stage. Unit testing is the responsibility of developers because it is done in the development phase by the developers, and it is the developers that create unit tests for their production-ready code.

Software developers can approach unit testing in two different ways, either writing their code before writing unit tests or before writing their actual code, where the developer first creates a failing unit test. The second approach is called **Test-Driven Development** (**TDD**).

When exploring the **Software Development Life Cycle** (**SDLC**) for software engineering and development, unit testing comes under development since it is the responsibility of the developers and serves as the base and first level of testing to ensure bug-free and defect-free software.

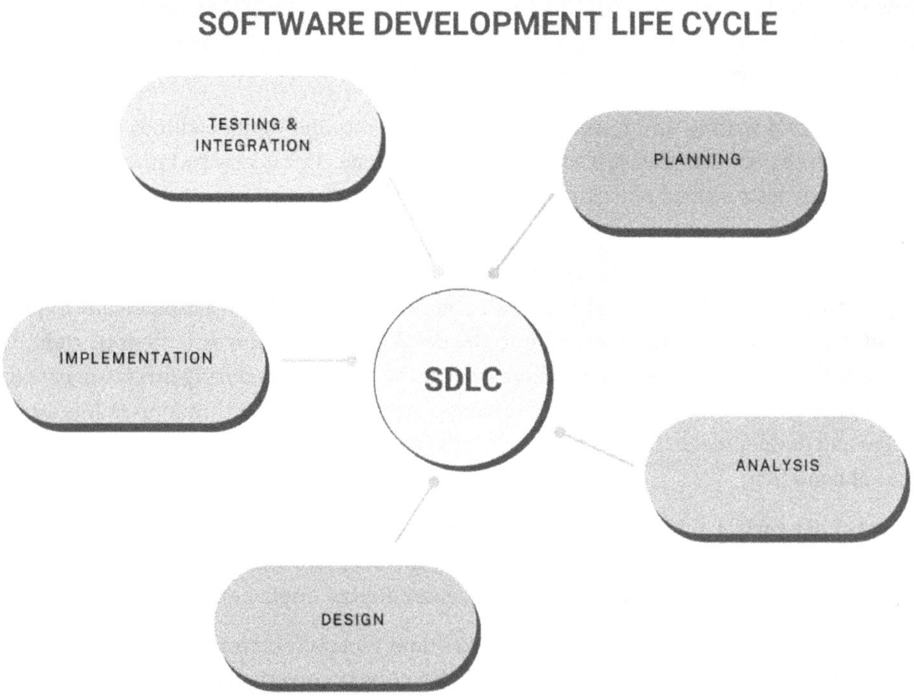

SOFTWARE DEVELOPMENT LIFE CYCLE

Figure 9.1 – SDLC

In the next section, we will examine different best practices to create and manage unit tests.

The importance and benefits of unit testing

The importance of implementing unit testing in your development pipeline cannot be overemphasized. It has been proven to have tremendous benefits. It has also helped in detecting errors early in the development phase. We will go through some of the major benefits and the importance of unit testing in this section.

Unit testing is used to design robust and enterprise-ready software components that help maintain code and eliminate issues in code units. Finding and fixing bugs during development is important compared to discovering them in production. Unit testing helps in fixing these errors early in the development phases.

It is an integral part of the agile software development process. During the build process and deployment, the unit test suite is automated to run and generate reports. If any of the unit tests fail, then the QA team should not accept that build for verification and it should be returned to the development team for more checking and validation.

Unit testing saves the QA and software testing teams a lot of time. If there is a proper standard and a well-configured, automated software testing pipeline for the enterprise application, errors and defects will be caught during development automatically.

Developers may avoid writing unit tests due to time constraints and tight deadlines. In most cases, they may opt to write poor unit tests just to have a 100% pass rate. This is very bad practice; it is better to avoid writing any tests instead of writing bad unit tests just to pass.

Here are some of the benefits of writing good unit test cases:

- **Improved code quality**: The quality of code shipped to production is automatically improved if unit testing is strictly implemented into the development pipeline and it's done right. Unit testing is the earliest form of testing; therefore, any bugs identified during this testing are fixed before they are sent to the integration testing phase. The result of this approach is a robust design and development as developers write test cases by understanding the specifications and tasks at hand first.

- **Detects bugs early**: Unit testing is the first level of testing in software development. Therefore, it helps in identifying and fixing bugs early. This includes flaws, missing parts in the software requirements and specifications, or bugs in the developers' implementation.

- **Saves development time**: Code completion takes time, especially if proper software development practices are in use. Therefore, when there are fewer bugs and errors in the system due to the effectiveness of unit testing, the overall development time is reduced.

- **Easy debugging process**: Unit testing helps in simplifying the testing and debugging process of an enterprise-ready application because if the test fails at any stage, the code needs to be debugged; otherwise, the process can continue without any obstacles.

- **Lower cost**: The cost of development, including development time, is drastically reduced when bugs are detected and resolved during development rather than during production. Without this testing, if the same bugs are detected at a later stage after the code integration, they become more difficult to trace and resolve, thereby making the development more costly and last longer.

The benefits and importance of unit testing are endless. Therefore, it's a good practice to adopt and implement it into your enterprise-ready Vue.js 3 application development pipeline. In the next section, we will explore the best practices for creating unit test cases.

Best practices in unit test creation

When creating unit test cases, you should follow the best practices to produce consistent unit test cases to test every possible case properly. Consider the following points to create good test cases:

- **Arrange, Act, and Assert (AAA)**
- Write deterministic tests

- Write good test names and descriptions
- Write tests before or during development (TDD)
- Leverage automated tests
- Using mocks and stubs

Let's understand each of these points in more detail in the next subsections.

Arrange, act, and assert

When structuring your unit test suite for enterprise applications, following the AAA approach is recommended to improve readability and easy understanding of your unit test suite. It improves the test readability by giving it a logical flow. It can also be referred to as the **Given/When/Then** (**GWT**) strategy.

GWT is a semi-structured way of writing down test cases. These test cases can either be manually tested or automated using LambdaTest (https://www.lambdatest.com/automation-testing?fp_ref=solomon26).

You can use the AAA protocol to structure your unit test cases with the following steps:

1. **Arrange**: Arrange the setup and initialization for the test.
2. **Act**: Act on the unit for a given test.
3. **Assert**: Assert or verify the outcome.

The following code snippet shows a basic example of using the AAA style to create a simple unit test case:

```
it('test for positive number', async () => {
  // Arrange
  const positive = 6;
  // Act
  const answer = Math.abs(positive);
  // Assert
  assert.equal(answer, positive);
});
```

The preceding snippet shows where to initialize variables and create the initial setup for the given test, then where we act on the given test, and lastly, where we assert the result of the acted-on test.

Write deterministic tests

A unit test should have consistent output whenever and wherever tested to verify the desired function. A deterministic test should have a consistent behavior every time the test is run provided the test code hasn't changed.

Inconsistencies in testing can be called flakiness in tests. If your test works or passes in development and fails with continuous integration or during QA testing, it hinders development and slows down progress.

Flakiness in tests can be avoided if deterministic test cases are written as it helps in understanding the output of every test case quickly and reduces hours of debugging for new team members.

Write good test names and descriptions

In software engineering, one of the best practices for writing clean code is to always have a good naming ability. As a developer, your variables, functions, methods, or classes should have good and descriptive names.

This best practice is also extended to writing test case names. It is important to have a clean and clear description of your test case to capture exactly when the test is supposed to implemented and the desired output.

For example, your test case names should describe the purpose of your test cases, as shown in the following examples:

```
describe("Test Names", () => {
  it("is a Vue instance", () => {});
  it("initializes with correct elements", () => {});
  it("test for positive number", async () => {});
  it('has a default message', async () => {});
  it('can change message', async () => {});
)};
```

Write tests before or during development (TDD)

As a professional developer, you need to embed the concept of TDD into your development process and workflow.

TDD is a software development process that enhances our test cases and software code in parallel.

The concept of TDD contradicts the traditional development process because in TDD, developers have to first write test code before writing the actual software code to pass the test case written. This approach ensures that when production code is written, it always complements the test code.

Additionally, **Behavior-Driven Development (BDD)** is another popular testing approach. This approach works well in rapid development settings and encourages more team collaboration to build a shared understanding of the problem.

Regardless of what approach you decide to adopt in your project, you can still integrate continuous integration into your development pipeline to automate your software testing processes.

Using mocks and stubs

When creating test cases, you might be tempted to perform operations on the actual code. For instance, if you made an API call to an external API, you might want to make such calls during testing, to make sure everything works as planned. But that wouldn't be considered best practice. What you can do is use the mock and stub features of any test framework.

A stub is a dummy piece of code that lets the test run without worrying about what happens to it, while a mock is a dummy piece of code that you verify is called correctly as part of the test. In short, they are substitutes for real, working code.

The beauty of this is that you can set them up and use them to test and verify your actual code works properly without making any expensive API calls or carrying out database manipulation.

Leverage automation testing

As a developer, integrating automated testing into your workflow saves a lot of time when compared to manually executing your test case every time before deployment.

You can use different automated testing frameworks to set it up, but in this chapter, we will see how to automate unit testing with Selenium Cloud Grid (`https://www.lambdatest.com/selenium-grid-online?fp_ref=solomon26`).

Before we delve into automating our test cases, let's explore how to write a clean and proper unit test with JavaScript.

JavaScript unit testing

As a developer, writing unit tests for your task is your responsibility. It should be part of your day-to-day activities as you code. In JavaScript, you can write unit tests the same way you write your real code with the use of different testing libraries.

With these testing libraries, testing the functionalities and features of your project becomes very easy because the libraries include different assertion methods to carry out your checks.

Let's explore some of the most popular JavaScript testing frameworks you can use to write your unit tests, integration tests, and even end-to-end tests.

Popular JavaScript testing frameworks

Various frameworks are helpful for unit testing in JavaScript. They are as follows:

- Jest
- Mocha
- Jasmine
- Cypress
- Vitest.js

Let's explore these frameworks in more detail in the next subsections.

Jest

Jest is one of the most popular testing frameworks for JavaScript. It was designed to mostly work with React and React Native-based applications. It is open source and easy to get started with. Jest reduces the extensive time-consuming configuration needed to run software testing in the frontend with JavaScript.

It is an assertion library for JavaScript that runs on Node.js and the browser. Jest can be configured to work with any test runner and unit testing framework, such as Mocha or Jasmine.

The growth statistics of the Jest library according to GitHub, as of the time of writing, include more than **40k GitHub stars** and about **6.3 million GitHub usage**, accumulating a total of **6.4 million** points, making Jest among the most popular testing frameworks.

Mocha

Mocha is a server-side and client-side testing framework for JavaScript and Node.js. The key features of Mocha are simplicity, flexibility, and fun. It makes asynchronous testing in JavaScript easy and fun. Mocha is designed to run serially, allowing for flexible and accurate test reporting and coverage.

The growth statistics of the Mocha library according to GitHub, as of the time of writing, include more than **21.6k GitHub stars** and about **1.6 million GitHub usage**, accumulating a total of **1.66 million**, points, making Mocha a very popular testing framework.

Jasmine

Jasmine is a popular JavaScript BDD framework for unit testing JavaScript applications. It combines the power of speed and support for Node.js and the browser to become a robust testing framework for BDD.

The growth statistics of the Jasmine library according to GitHub, as of the time of writing, include more than **15.4k GitHub stars** and about **2.4 million GitHub usage**, accumulating a total of **2.5 million** points, making Jasmine among the most popular testing frameworks.

Cypress

Cypress is an end-to-end JavaScript-based testing framework that changes the way developers approach software testing. It is built on top of Mocha, making asynchronous testing simple and convenient. In Cypress, unit tests can be configured to execute without even having to run the web server.

This feature makes Cypress the ideal tool for testing a JavaScript/TypeScript library that is meant to be used in the browser, and setting up automated testing in your enterprise project is a breeze.

The growth statistics of the Cypress library according to GitHub, as of the time of writing, include more than **40.2k GitHub stars** and about **535k GitHub usage**, accumulating a total of **575k** points, making Cypress among the most popular testing frameworks.

Vitest

Vitest is a blazing-fast unit test framework powered by Vite. It is a Vite-native unit test framework comprising Vite reusable configs, transformers, resolvers, and plugins. It is also Jest compatible and uses ESM, TypeScript, and JSX out of the box.

It's fairly new and has gained popularity among developers using Vue.js and the Vite CLI. The growth statistics of the Vitest library according to GitHub, as of the time of writing, include more than **6.4k GitHub stars** and about **24.3k GitHub usage**, accumulating a total of **30.7k** points.

In the next section, we will explore how to write your unit tests with JavaScript and how to run your tests manually. Additionally, we will explore how to automate your testing suite during the deployment pipeline.

Unit testing a Vue.js 3 app

In the previous chapter, we created a Pinterest application using Strapi for the backend and Vue.js 3 for the frontend.

Previously, we added internationalization, structured the project, and built out a complete Pinterest clone. In this section, we will continue by using the official project we created for this book to set up unit testing, resulting in a full-blown enterprise-ready Pinterest clone application.

You can clone the project from this repository, `https://github.com/PacktPublishing/Architecting-Vue.js-3-Enterprise-Ready-Web-Applications`, to jump right in.

In the previous chapter, we set up basic unit testing using Vitest and demonstrated how to implement basic unit testing with a helper file.

In this chapter, we will explore more advanced ways of testing the units of the Pinterest clone application we are using as the example throughout this book.

What to test

In the previous chapter, we explored in detail what to test when setting up your testing suites. In this section, we will examine our demo software under test and specify what should be unit tested.

In general, there are two things we could test for in Vue.js components: presentation and (optionally) behavior.

Presentation

When fetching data using Apollo Client, components can be in either the `loading`, `success`, or `error` state. For each of these states, it's a good idea to test that the presentation is what we intend it to be.

For example, consider having a component that presents details about a specific photo (performs a `GET_PIN` query) from our Pinterest clone application.

We could have a simple component such as the following:

```
<template>
 <div v-if="status === 'loading'">Loading photo...</div>
 <div v-else-if="status === 'error'">An error occurred
 </div>
 <div v-else>
   <img src="/src/assets/kunal-img.jpg" alt=""
     data-testid="pin"
     class="w-full h-full object-cover rounded-2xl" />
 </div>
</template>
<script setup>
import Button from '../atoms/Button.vue';
defineProps({
 photo: { type: Object, default: () => { } },
 status: { type: String, default: "Loading" }
});
</script>
```

In this scenario, we will likely want to test the following:

- **Loading**: How the component renders when it's fetching the pin (photo)
- **Success**: How the component renders after it's successfully fetched the pin (photo)
- **Error**: How the component renders if it was unable to fetch the pin (photo)

To demonstrate this, let's implement a unit test against each of the states mentioned in the preceding points. We will be using the `Pin(Photo)` component:

```
import { render } from '@testing-library/vue';
import Card from '../../components/molecules/Card.vue';
test('displays a card with success status', async () => {
  const { getByTestId } = render(Card, {
    props: {
      status: 'success',
    },
  });
  const card = getByTestId('pin');
  expect(card).toBeDefined();
});
test('displays a card with error status', async () => {
  const { getByText } = render(Card, {
    props: {
      status: 'error',
    },
  });
  const card = getByText('An error occurred');
  expect(card.textContent).toEqual('An error occurred');
});
test('displays a card with loading status', async () => {
  const { getByText } = render(Card, {
    props: {
      status: 'loading',
    },
  });
  const card = getByText('Loading photo...');
```

```
    expect(card.textContent).toEqual('Loading photo...');
});
```

Behavior (optional)

We may also choose to place behavior in our Vue.js components. In client-side architecture, we call this interaction logic—a form of decision-making logic executed after the user interacts with the page somehow—such as a key press or a button click.

You can also unit test the behaviors of a single component by testing different actions that are performed on the component and making sure the component reacts to it accordingly.

Let's test this `Photo` component's `click` event to be sure it responds to the appropriate action:

```
test('clicks a create pin button', async () => {
  const { getByTestId, emitted } = render(Card);
  await fireEvent.click(getByTestId('create_pin'));
  expect(emitted()).toHaveProperty('click');
});
```

You can follow the preceding sample code snippet to write unit tests for all the components you have created in your enterprise Vue.js 3 application.

In the repository of this chapter (https://github.com/PacktPublishing/Architecting-Vue.js-3-Enterprise-Ready-Web-Applications/tree/chapter-9), we have created different components and also written the unit test for them. You can clone the repository here.

In the next section, we will run the unit testing manually and how to automate the process using LambdaTest Cloud Grid. (https://www.lambdatest.com/automation-testing/?fp_ref=solomon26)

Running unit tests manually

To run your test, type the following command into your root terminal:

```
npm run test:unit
// or
yarn test:unit
```

After successfully running the test, you should be greeted with green passes for your test, as in the following screenshot:

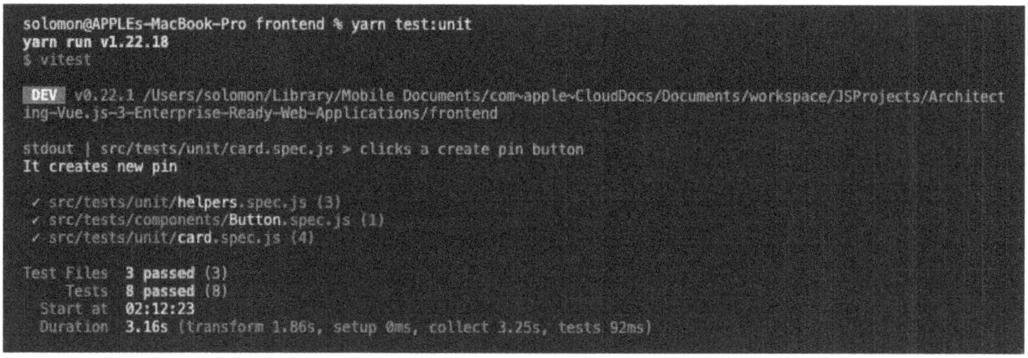

Figure 9.2 – Unit test passing sample

In this section, we explored best practices in unit testing a Vue.js 3 component. We discussed, most importantly, what to test and how to implement unit testing in Vue.js 3. We demonstrated how to unit test a Vue.js app using Vitest (`https://vitest.dev/`) and the Vue.js testing library. (`https://github.com/testing-library/vue-testing-library`)

Summary

This chapter explored everything related to unit testing, including how to unit test a Vue.js 3 component and the component and page methods. We also learned about unit testing tools such as Jest, Mocha, and Vitest and how to use them to effectively unit test an enterprise project.

In this chapter, we explored the benefits, importance, and best practices in writing and executing effective unit testing strategies. We also learned how to write unit test cases based on the presentation and behavior of the different units of the software under test.

This chapter shows you how to create, implement, and run your unit test cases manually during the build process and deployment.

In the next chapter, we will explore everything related to integration testing. We will cover in depth how to perform an integration test on a Vue.js 3 component and pages. We will also learn about integration testing tools such as Vue.js Testing Library and how to use them to test an enterprise project effectively.

After successfully finishing the test, you should be greeted with green passes for your tests in the following screenshot:

Figure 9.7 – Unit test pass on console

Integration Testing in Vue.js 3

In the previous chapter, we learned about everything related to unit testing. We learned how to unit-test a Vue.js 3 component and the components and pages' methods. We also learned about unit testing tools such as Jest and Mocha and used them to effectively unit-test an enterprise project.

In this chapter, we will explore everything related to integration testing. We will learn in depth how to perform an integration test on a Vue.js 3 component and pages. We will also learn about integration testing tools such as Vue Test Library (`https://github.com/testing-library/vue-testing-library`) and how to use them to test an enterprise project effectively.

We will cover the following key topics in this chapter:

- Introduction to integration testing
- What is integration testing?
- Importance of integration testing
- Benefits of integration testing
- Best practices when creating integration tests
- JavaScript integration testing
- Testing a basic Vue app

Technical requirements

To get started with this chapter, I recommend you read through *Chapter 9*, *Best Practices in Unit Testing*, where we explored the benefits, importance, and best practices in writing and executing effective unit testing strategies. We will rely heavily on the knowledge acquired from that chapter in this one to learn about Vue 3 enterprise integration testing.

All the code files for this chapter can be found at `https://github.com/PacktPublishing/Architecting-Vue.js-3-Enterprise-Ready-Web-Applications/tree/chapter-10`.

Introduction to integration testing

Engineers in a team develop applications in isolation, and after development and unit testing each unit during development, the next phase in the software testing phase is integration testing. This form of testing involves testing the modules/components when they are combined/integrated to make sure that they work according to the requirement.

It is a type of testing where the units of software modules are integrated logically and tested completely as a group.

In this section, we will examine the definition of integration testing and explore how developers can create and manage integration test cases within their enterprise-ready Vue 3 application.

What is integration testing?

Integration testing is a type of testing in which the different units, modules, or components of the **software under test** (**SUT**) are combined and tested as a single entity. In addition, these modules or units are independent of the developers or team and can be coded by different programmers.

It is also known as component testing or **integration and testing** (**I&T**).

Integration testing is the first stage of combining individual modules to form components or combined entities, and it is aimed at testing the interfaces between the modules to expose any defects that may arise when these components are integrated and interact with each other.

In the next section, we will examine the importance and benefits of integration testing for agile and enterprise-level teams.

Importance of integration testing

Integration testing is a critical phase of the software testing process. It is the process of testing the interactions and interfaces between different components or modules of a system. It is important for several reasons:

- It helps to ensure that different system components work together seamlessly and as expected
- It helps to identify and resolve conflicts between different system components
- It helps to identify and resolve bugs that may not have been uncovered during unit testing
- It helps to identify and resolve performance bottlenecks that may not have been uncovered during unit testing
- It helps ensure the system can handle the expected load and usage patterns
- It helps ensure the system is secure and can protect sensitive data
- It helps to ensure that the system can be integrated with other systems or external components

- It helps to ensure that the system meets the requirements and specifications defined in the design phase

- It helps to identify and resolve issues with data flow and data integrity between different system components

- It can help to identify and resolve issues with **user interfaces** (**UIs**) and user interactions between different system components

- It can help to identify and resolve issues with third-party APIs and services that the system may need to interact with

- It can help ensure the system can handle different environments and configurations, such as different operating systems or different hardware configurations

- It can help ensure the system is compatible with other systems it may need to interact with, such as databases or external services

Overall, integration testing is an essential step in the software development process that helps to ensure the quality and reliability of the final product. It is important to conduct integration testing early in the development process so that any issues can be identified and resolved as soon as possible.

It's also important to note that integration testing is not limited to testing between the different components of the system. It also includes testing the system as a whole with other external systems it will integrate with. It's good practice to test the integration of the system with other systems before releasing it to production.

In the next section, let's look at some of the benefits of integration testing.

Benefits of integration testing

Integration testing is particularly important for enterprise software systems, as they are often complex, multi-faceted systems that need to integrate with other systems and handle large amounts of data and transactions. Some of the benefits of integration testing for enterprise software include the following.

Improved system reliability

By testing the interactions and interfaces between different components of the system, integration testing helps to ensure that the system is reliable and can handle the expected load and usage patterns. This can help to reduce downtime and improve the system's overall performance.

Reduced risk of data loss

Enterprise systems often handle large amounts of sensitive data, and integration testing can help ensure that data is properly protected and that data integrity is maintained between different system components.

Increased scalability

Integration testing can help to identify and resolve performance bottlenecks that may not have been uncovered during unit testing, making it easier to scale the system as needed.

Improved security

Integration testing can help to ensure that the system is secure and can protect sensitive data, reducing the risk of data breaches and other security incidents.

Better integration with other systems

Testing a system's integration with other systems before it is released to production can help to ensure that the system is compatible and able to communicate with other systems that it may need to interact with.

Better compliance

By conducting integration testing, you can ensure that the system meets the requirements and specifications defined in the design phase, which can help to ensure compliance with industry standards and regulations.

Better customer satisfaction and ROI

By ensuring that the system is reliable, secure, and easy to use, integration testing can help to improve customer satisfaction and increase system adoption. Ensuring that all the system components are working seamlessly and without bugs can help to reduce development costs, improve the system's performance, and increase the overall **return on investment (ROI)** of the system.

Improved testing efficiency

Integration testing can identify issues early in the development process, which can help to reduce the overall time and cost of testing.

Improved software quality

By thoroughly testing the interactions and interfaces between different components of the system, integration testing can help to ensure that the software is of high quality and free of defects.

Improved team collaboration

Integration testing often involves collaboration between different teams, such as development, testing, and operations teams. This can help to improve communication and collaboration between teams and ensure that the system is developed and tested to meet the needs of all stakeholders.

Improved documentation

The integration testing process can lead to better documentation of test cases and test results, which can be used to improve the system and for future reference.

Improved business continuity

By ensuring that the system is reliable and can handle the expected load and usage patterns, integration testing can help to ensure that the system can continue to operate in case of unexpected events, such as hardware failures or power outages, which can help to improve business continuity and minimize disruption.

Improved data governance

By thoroughly testing the interactions and interfaces between different system components, integration testing can help ensure that data is properly protected and that data integrity is maintained between different components of the system, which can help improve data governance and compliance with data protection regulations.

In this section, we explored the benefits of integration testing to enterprise projects and how it improves development teams' workflows. In the next section, we will learn about some of the best practices to put in place during integration testing.

Best practices when creating integration tests

Now let's look are some best practices for creating effective integration tests.

Starting early

Integration testing should be started as early as possible in the development process, ideally during the design phase. This will help to identify and resolve issues early on and ensure that the system is developed and tested to meet the needs of all stakeholders.

Defining clear objectives

Before starting integration testing, it's important to define the objectives of the testing clearly. This includes identifying what the system is supposed to do, the system's interfaces and interactions, and the expected outcomes.

Creating a test plan

Create a test plan that outlines the scope of the testing, the test cases to be executed, and the resources and tools needed. The test plan should also include a schedule for testing, including when testing will be completed and when the results will be reviewed.

Using a modular approach

Divide the system into smaller, more manageable modules and test them separately. This will help to identify and resolve issues more quickly and efficiently.

Using automated testing

Automated testing can help to improve testing efficiency and reduce the time and cost of testing. Automated testing can also be used to test the system under different conditions, such as different environments and configurations.

Testing the system as a whole

Test the system as a whole with the other external systems it will integrate with; this will ensure that the system is compatible and able to communicate with other systems that it may need to interact with.

Testing for security vulnerabilities

Test the system for security vulnerabilities, such as SQL injection or cross-site scripting attacks. This will help ensure the system is secure and can protect sensitive data.

Testing for performance bottlenecks

Test the system for performance bottlenecks and identify and resolve any issues that may arise. This will help ensure the system can handle the expected load and usage patterns.

Documenting and reviewing the results

Document the results of the testing, including any issues that were identified and how they were resolved. Review the results and use them to improve the system and the testing process.

Continuously testing and monitoring

Continuously test and monitor the system after it's been released to production; this will help to ensure that the system is reliable and that any issues are identified and resolved quickly.

In this section, we explored the best practices to put in place when making use of integration testing into an enterprise. In the next section, we will learn how to implement integration testing within the demo project using JavaScript.

JavaScript integration testing

There are several tools that can be used to perform integration testing in JavaScript. Some popular choices include the following.

Mocha

Mocha is a widely used JavaScript testing framework well suited for integration testing. It is highly customizable and can be used in conjunction with other libraries, such as Chai and Sinon, to perform various types of testing.

Cypress

Cypress is a JavaScript-based end-to-end testing framework that can be used for integration testing. It allows developers to test the entire flow of an application from the user's perspective, and it has built-in support for real-time debugging, automatic waiting, and time-traveling.

TestCafe

TestCafe is an end-to-end testing tool that runs on top of Node.js. It allows you to run tests in a real browser and is easy to set up and use. It also offers the ability to test the UI of your application, which is useful in integration testing.

Selenium

Selenium is a browser automation tool that can be used for integration tests for web applications. Selenium WebDriver allows you to interact with web browsers and perform tasks such as clicking buttons, filling out forms, and navigating through pages.

Vue Test Utils

Vue Test Utils is an official testing library provided by the Vue.js team. It is a lightweight library that provides a set of utilities for testing Vue components. It can be used in conjunction with other testing frameworks, such as Jest or Mocha, to perform integration testing.

Avoriaz

Avoriaz is a testing library specifically designed for Vue.js components. It provides a set of tools for testing Vue components and allows you to easily mount and interact with your components in a testing environment.

Vue Testing Library

This is a library built for testing Vue 3 applications. It provides a set of utilities for testing Vue 3 components and allows you to easily mount and interact with your components in a testing environment.

Nightwatch

Nightwatch is an automated testing framework for web applications and websites. It can be used for integration tests for Vue 3 applications, and it allows you to write integration tests that simulate user interactions with an application.

As with any other tool, you can pick the one that best suits your needs and the structure of your application. Keep in mind that some tools may be better suited for certain types of testing or certain types of applications.

These are some of the most popular tools available; you can use any depending on your needs and what you are trying to test.

It's also important to note that most of these tools can be integrated with other libraries and frameworks to extend their functionality.

In the next section, we will explore how to write your integration test with JavaScript and how to run your test manually. We will also explore how to automate your testing suite during a deployment pipeline.

Testing a basic Vue app

In the previous chapter, we created a Pinterest application using Strapi for the backend and Vue 3 for the frontend.

Also, we added internationalization, structured the project, implemented unit and integration testing, and built a complete Pinterest clone. In this section, we will continue by using the official project we created for this book to set up integration testing to make up a full-blown enterprise-ready Pinterest clone application.

You can clone the project from the GitHub link mentioned in the *Technical requirements* section.

In this chapter, we will explore more advanced ways of implementing integration testing within the Pinterest clone application we use throughout this book.

Writing a basic integration test

First, we will start by installing the new testing library for Vue 3. Since we are using Vite in the project, we will also install the Vitest library for our test runner. You can read more about the new Vue 3 test library at `https://vitest.dev/guide/`.

Installing the test library

As of the time of writing, `@testing-library/vue` (https://github.com/testing-library/vue-testing-library) and `vitest` (https://vitest.dev/guide/) are recommended for integration testing, and we will install the test libraries using the following command:

```
npm install -D vitest happy-dom @testing-library/vue
```

Next, open your `vite.config.js` file and add the following configuration. Note that happy-dom is a JavaScript implementation of a web browser without its graphical UI:

```
import { defineConfig } from 'vite';
import vue from '@vitejs/plugin-vue';
// https://vitejs.dev/config/
export default defineConfig({
  plugins: [vue()],

  test: {
    environment: 'happy-dom',
    globals: true,
  },
});
```

In the preceding code, we configured the testing library to accommodate both unit and integration testing using the Vite library. This means that if we want to run unit and integration testing, we only need to use a single command, as shown when running the test.

Next, let's create a simple integration test.

Creating a basic component test

Here is a simple integration test snippet to demonstrate. This is a general integration testing example that tests the `Button` component inside the Pinterest demo application:

```
import { render } from '@testing-library/vue';
import Button from '../../components/atoms/Button.vue';
test('mounted a button with custom label', async () => {
```

```
// The render method returns a collection of utilities to
// query your component.
const { getByText } = render(Button, {
  props: {
    label: 'Test',
  },
});
// getByText returns the first matching node for the
// provided text, and  if button is rendered with Label from
// props
const button = getByText('Test');
});
```

The test simply renders Button with a custom label value and also checks whether we can retrieve the custom label added during the rendering process.

Running the test

Running the test will result in four passed test cases, including the unit test we created earlier. Here's a command to run this test sample:

```bash
yarn test
```

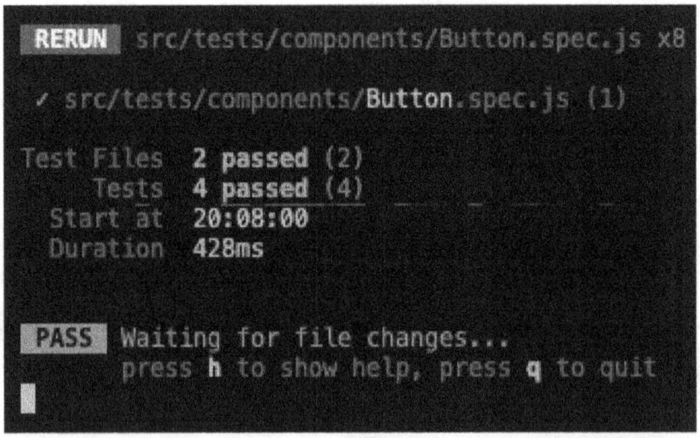

Figure 10.1 – A screenshot of the general test result

In this section, we have demonstrated how to configure and structure integration testing, also known as component testing, with Vue 3 using the latest Vitest testing library for Vue 3. We have also written basic component tests to help us understand the process of writing integration tests. In the next section, we will explore testing integrated components with Vue 3 using Vitest.

Testing integrated components

In the previous example, we tested a simple Button component to make it render properly with the required properties. In this section, we will test a completely integrated component that combines different individual components. Let's get started with the following steps:

1. Create a file inside the tests/components directory called HomeOverivew.vue since we want to test the integration of the home page.

2. Open the file and add the following testing code or clone the repository using the GitHub link mentioned in the *Technical requirements* section:

```
import { fireEvent, render } from "@testing-library/vue";
import { describe, expect, it } from "vitest";
import HomeOverview from "../../components/templates/
HomeOverview.vue";

describe("HomeOverview.vue", () => {
  it("renders component", async () => {
    const { getByText } = render(HomeOverview);
    getByText("Home");
  });

  it("creates pin on button click", async () => {
    const { getByTestId, emitted } =
      render(HomeOverview);
      await fireEvent.click(getByTestId("create"));
      // Fireevent is from "@testing-library/vue" for
      // calling different events such as click
    expect(emitted()).toHaveProperty("click");
  });

  it("dismisses notification", async () => {
    const { getByTestId, emitted } =
      render(HomeOverview);
```

```
    await fireEvent.click(getByTestId("dismissed"));
    expect(emitted()).toHaveProperty("click");
  });

  it("displays first 14 pins", async () => {
    const { getAllByText } = render(HomeOverview);
    const card = getAllByText("Quick save and organize
                              later");
    expect(card.length).toBe(14);
  });

  it("renders Search component", async () => {
    const { getByTestId } = render(HomeOverview);
    getByTestId("search");
  });
});
```

In each test case, we are testing components that were added to the HomeOverview component to demonstrate how we can use integration testing to test integrated components as one.

3. Next, make sure the HomeOverivew page has been rendered correctly before testing for other test cases:

```
it("renders component", async () => {
  const { getByText } = render(HomeOverview);
  getByText("Home");
});
```

4. Next, test whether the Button component renders correctly and also check whether we can perform some actions with it. For instance, when the button is clicked, the app is supposed to create a new pin with photos. We will test to make sure that this functionality is implemented correctly even after integration:

```
it("creates pin on button click", async () => {
  const { getByTestId, emitted } =
    render(HomeOverview);
  await fireEvent.click(getByTestId("create"));
  expect(emitted()).toHaveProperty("click");
});
```

5. Next, test the notification display component found inside the `Header` component. We are testing it to make sure the users can dismiss notifications, and that it is also rendered correctly both inside the `Header` component and the `HomeOverview` component:

```
it("dismisses notification", async () => {
    const { getByTestId, emitted } =
        render(HomeOverview);
    await fireEvent.click(getByTestId("dismissed"));
    expect(emitted()).toHaveProperty("click");
});
```

6. Next, also test the `Card` component to make sure that it displays the total amount of pins on the home page, and also that it renders the components correctly:

```
it("displays first 14 pins", async () => {
    const { getAllByText } = render(HomeOverview);
    const card = getAllByText("Quick save and organize
                                later");
    expect(card.length).toBe(14);
});
```

7. Lastly, we will test the `Search` component to make sure it was properly rendered for users and is available for users to search for pins:

```
it("renders Search component", async () => {
    const { getByTestId } = render(HomeOverview);
    getByTestId("search");
});
```

8. Now, let's run the test by running the following command in our root terminal:

```bash
yarn test:component
```

After successfully performing running the test, you should be greeted with green passes for your test, as in the following screenshot:

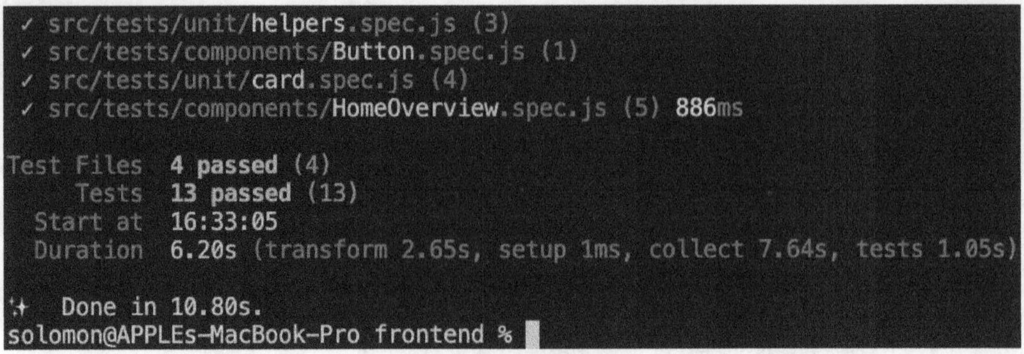

```
✓ src/tests/unit/helpers.spec.js (3)
✓ src/tests/components/Button.spec.js (1)
✓ src/tests/unit/card.spec.js (4)
✓ src/tests/components/HomeOverview.spec.js (5) 886ms

Test Files  4 passed (4)
     Tests  13 passed (13)
  Start at  16:33:05
  Duration  6.20s (transform 2.65s, setup 1ms, collect 7.64s, tests 1.05s)

✦+  Done in 10.80s.
solomon@APPLEs-MacBook-Pro frontend %
```

Figure 10.2 – A screenshot of the integration test result

With all these test cases successful, we can easily see how integration testing helps developers to test out integrated and combined components instead of testing these components in isolation.

Summary

This chapter explored everything related to integration testing. You also learned about integration testing tools such as Cypress, Mocha, and Vue Testing Library and used Vue Testing Library to effectively test an enterprise project.

In this chapter, we explored the benefits, importance, and best practices in writing and executing effective integration testing strategies. In addition, you also learned how to write integration test cases.

In the next chapter, you will learn about everything related to end-to-end testing. You will learn in depth how to perform end-to-end testing on a Vue.js 3 component and pages. In addition, you will also learn about end-to-end testing tools such as Cypress and Puppeteer and how to use them to test an enterprise project effectively from end to end.

11

Industry-Standard End-to-End Testing

In the previous chapter, we learned about everything related to **integration testing**. We learned in depth how to perform an integration test on a Vue.js 3 component and pages. Additionally, we learned about integration testing tools such as Vue Test Library and how to use them to test an enterprise project effectively.

In this chapter, we will explore everything related to **end-to-end testing** (**E2E**). We will learn how to perform E2E testing on a Vue.js 3 component and pages. In addition, we will learn about E2E testing tools, such as Cypress and Puppeteer, and how to use them to perform an E2E test in an enterprise project effectively.

In this chapter, we will cover the following key topics:

- Introduction to E2E testing

- What is E2E testing?

- Importance of E2E testing

- Benefits of E2E testing

- Best practices in creating E2E test

- JavaScript E2E testing

- E2E-testing a Vue app

Technical requirements

To get started with this chapter, I recommend you to read through *Chapter 10, Integration Testing in Vue.js 3*, where we explored everything related to integration testing. Also, we learned in depth how to perform an integration test on a Vue.js 3 component and pages. We will rely heavily on the knowledge from that chapter in this chapter to learn about Vue.js 3 enterprise E2E testing.

All the code files for this chapter can be found at `https://github.com/PacktPublishing/Architecting-Vue.js-3-Enterprise-Ready-Web-Applications/tree/chapter-11`.

Introduction to E2E testing

E2E is a complex testing process that assesses the working order of a complex application from start to finish. Additionally, you can use E2E testing to work through a complete application exactly how you intend your end users to use the product and discover any bugs before pushing the code to production for real users.

In most organizations, E2E testing is a standard used to access the complete features of an application after developing them in isolation with different developers in your team.

E2E testing in large teams is possible by having a central repository system that is used to build and combine the code base. Next, E2E testing runs through the completed features and ensures they work as intended before approving and pushing them to the production stage.

In this section, we will examine the definition of E2E testing and explore how developers can create and manage E2E test cases within their enterprise-ready Vue.js 3 application.

What is E2E testing?

E2E testing is a methodology that assesses the working order of a complex product in a start-to-finish process. It ensures that the application behaves as expected and that the data is maintained and flows in the same direction as expected for each task and process.

This testing aims to replicate real user scenarios to validate the system for integration and data integrity. The test goes through every operation the application can perform, including communicating with external devices, to make sure the actions of the end users are replicated and tested.

In the next section, we will examine the importance and benefits of E2E testing for agile and enterprise-level teams.

Importance of E2E testing

E2E testing makes it simpler to catch problems before releasing the software to end users. Additionally, it helps managers prioritize tasks in the development backlog by identifying the importance of a workflow to end users.

Moreover, for enterprise-level applications, E2E testing improves the user experience for multiple application interfaces such as web, desktop, and mobile apps because user expectations are the basis for the test cases.

E2E testing has been widely adopted because it helps reduce the overall cost of building and maintaining software by decreasing the time it takes to test software.

It helps a team expand its test coverage by adding more detailed test cases than other testing methods such as unit and functional testing.

It also ensures that the application performs correctly by running the test cases based on an end user's behavior.

Benefits of E2E testing

E2E testing is a very complex form of testing as it tests the behaviors of the end user. Therefore, it is crucial to follow the practices outlined in the following subsections to ensure smooth testing.

Reducing risks

The E2E testing process ensures rigorous testing of the software under test at the end of each iteration, thereby reducing the risk of future failures in production.

Consistent user experience

E2E testing involves testing the frontend. It ensures that the software under test provides a user experience that works across multiple devices, platforms, and environments.

Reducing cost and time

The cost and the number of times you need to test enterprise applications can be reduced by automating your E2E tests This reduces the amount of time and money it takes to maintain and improve the application.

Increasing confidence

E2E testing not only ensures that the application functions correctly but also increases confidence in its performance because it has been tested across multiple devices and platforms.

Less repetitive efforts

It reduces the chances of frequent breakdowns and, ultimately, reduces repetitive testing efforts due to more thorough and rigorous E2E testing.

Ensuring the correctness of the application

E2E is an essential software testing methodology because it validates an application at all layers – data, business rules, integration, and presentation. Therefore, it helps to ensure the correctness and health of the application.

In this section, we discussed the importance and benefits of E2E testing, illustrating why companies need to integrate it into their development pipelines. In the next section, we will explore the best practices involved in creating E2E testing.

Best practices in E2E test creation

E2E testing mimics the actions, activities, and experiences of a real user using the application. When creating E2E test cases, you should follow these best practices to produce consistent E2E test cases to test every possible case properly. Consider the following points to create good test cases.

Prioritizing the end use

When creating test cases, test like the user, and get into the mindset of someone using the app for the first time. Furthermore, ask and answer some of the user's questions such as *is it easy to find all the options? Are the features marked? Can users get what they want in fewer steps?*

Prioritizing the right aspects

It's important to prioritize what you're testing because this can easily become cumbersome and complex. Therefore, it's important to prioritize business-impacted features before going over other less important edge cases.

Making testing realistic

Sometimes, you want to make E2E testing a little realistic. In most cases, real users stop by to look at images, or pause and watch a few videos before moving on with their actions. E2E testing should mirror real-life interactions as much as possible.

Testing repeated user scenarios

E2E testing is very complex and requires time to test out all the possible edge cases completely. Avoid testing every possible edge case and focus only on the most common and important scenarios.

Error monitoring

E2E testing is a very complex process because it encompasses the walk-through of the whole application or, sometimes, only features that have been newly added. However, the complexity can be reduced by making sure many errors are resolved during coding before E2E testing.

Optimizing the testing environment

You can facilitate the testing process by creating an optimum test environment. Creating an optimum test environment allows for minimal setup when it's time to test and clear out data for your next test.

We have explored the best practices when it comes to implementing E2E testing and discussed a few points that you must consider when building E2E testing solutions. In the next section, we will learn more about different JavaScript E2E testing libraries.

JavaScript E2E testing

Various frameworks are helpful for unit testing in JavaScript. They are as follows:

- Selenium WebDriver
- Cypress
- Playwright
- Puppeteer
- Karma

Next, we will explore each of the libraries, discuss their popularities, similarities, and differences, and look at why you should choose any of these libraries for your E2E testing solution.

Selenium WebDriver

Selenium WebDriver is the most popular E2E testing software. It's a web framework that allows you to execute cross-browser tests by automating web-based application testing to verify that it performs as expected:

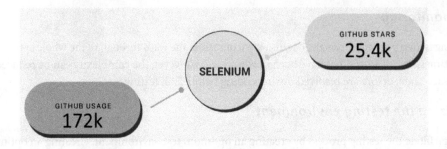

Figure 11.1 – A diagram showing Selenium statistics

Some of the growth statistics of the Selenium library at the time of writing, according to GitHub (`https://github.com/seleniumhq/selenium`), include more than **25.4k GitHub Stars** and about **172k GitHub Usage**, making Selenium among the most popular testing frameworks.

Cypress

Cypress is an E2E JavaScript-based testing framework that changes how developers approach software testing. It is a testing framework that does not use Selenium or WebDriver, making it faster and easy to set up for enterprise-level testing.

This feature makes Cypress the ideal tool for testing a JavaScript/TypeScript library that is meant to be used in the browser, and setting up automated testing with it in your enterprise project is a breeze:

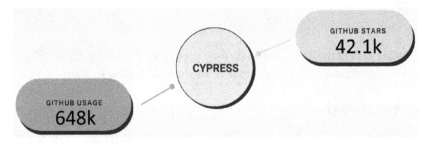

Figure 11.2 – A diagram showing Cypress statistics

Some of the growth statistics of the Cypress library at the time of writing, according to GitHub (`https://github.com/cypress-io/cypress`), include more than **42.1k GitHub Stars** and about **648k GitHub Usage**, making Cypress among the most popular testing frameworks.

Playwright

Playwright enables reliable E2E testing for modern web apps. It supports all modern rendering engines including Chromium, Webkit, and Firefox. Additionally, it supports cross-platform testing for Windows, Linux, and macOS including local, CI, headless, or headed testing. Most importantly, you can test with mobile web and with different programming languages:

Figure 11.3 – A diagram showing Playwright statistics

Some of the growth statistics of the Playwright library at the time of writing, according to GitHub (`https://github.com/microsoft/playwright`), include more than **46k GitHub Stars**, making Playwright among the most popular testing frameworks.

Puppeteer

Puppeteer is a Node.js library developed by Google that lets you control headless Chrome programmatically. You can automate the testing of your web applications, run testing in the browser, and see the results in real time on your Terminal session:

Figure 11.4 – A diagram showing Puppeteer statistics

Some of the growth statistics of the Puppeteer library at the time of writing, according to GitHub (`https://github.com/puppeteer/puppeteer`), include more than **81.3k GitHub Stars** and about **271k GitHub Usage**, making Puppeteer among the most popular testing frameworks.

Karma

Karma is an E2E testing framework that spawns a web server that executes source code against test code for each of the browsers connected. The results are displayed to the developers to see whether the test case failed or passed.

The Angular team created the Karma test library to fit their ever-changing testing requirements to make life easier:

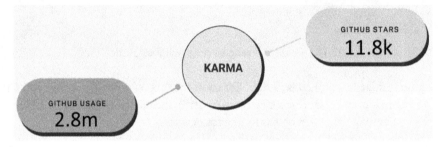

Figure 11.5 – A diagram showing Karma statistics

Some of the growth statistics of the Karma library at the time of writing, according to GitHub (`https://github.com/karma-runner/karma`), include more than **11.8k GitHub Stars** and about **2.8m GitHub Usage**, making Karma among the most popular testing frameworks.

In the next section, we will explore how to write your E2E test with JavaScript and how to run your test manually. Additionally, we will explore how to automate your testing suite during a deployment pipeline.

E2E testing a Vue.js 3 app

In the previous chapter, we created a Pinterest application using Strapi for the backend and Vue.js 3 for the frontend.

In the previous chapters, we added internationalization, structured the project, implemented unit and integration testing, and built out a complete Pinterest clone. In this section, we will continue by using the official project we created for this book to set up E2E testing to make up a full-blown enterprise-ready Pinterest clone application.

You can clone the project from `https://github.com/PacktPublishing/Architecting-Vue.js-3-Enterprise-Ready-Web-Applications` and jump right in.

In the previous chapter, we set up basic unit testing using Vitest and demonstrated how to implement basic unit testing with a helper file.

In this section, we will explore more advanced ways of implementing E2E testing on the Pinterest clone application we use throughout this book. We will be using Cypress for our E2E testing.

Setting up Cypress

Writing an E2E test is simpler than you think using some of the popular testing frameworks listed previously. In this section of the JavaScript E2E testing tutorial, we will use the Cypress framework to write E2E test cases.

We will write some E2E test cases to log a user in to our Pinterest clone application and check whether the user details are correct or not, but before that, let's install and configure Cypress.

Installing and configuring Cypress

You can set up a new project and install and configure Cypress by following this chapter. However, you can also clone the `chapter 11` folder from the official repository to follow along. At the time of writing, the following libraries will need to be installed:

```bash
yarn add cypress @cypress/vue @cypress/webpack-dev-server
```

Next, add the following script to your `package.json` file:

```bash
  "test:e2e": "cypress open --e2e",
  "test:e2e:ci": "cypress run --e2e"
```

The new script will allow you to run only the E2E test and see the outputs without running other test cases.

Lastly, create a `cypress.config.js` file in the root directory of your project and add the following configuration:

```
const { defineConfig } = require("cypress");
module.exports = defineConfig({
 component: {},
 env: {
   // HINT: here we read these keys from the .env file,
```

```
    // feel free to remove the items that you don't need
    baseUrl: "http://localhost:3000",
    apiUrl: "http://localhost:1337",
    email: "admin@test.com",
    password: "Admin111",
  },
  e2e: {
    supportFolder: false,
    supportFile: false,
    specPattern: "src/tests/e2e/**/*.spec.js",
    // eslint-disable-next-line no-unused-vars
    setupNodeEvents(on, config) {
      // implement node event listeners here
    },
    baseUrl: "http://localhost:3000",
  },
});
```

The variables inside env are optional and only contain variables specific to my environment variables; you should update this to reflect your environment variables.

Next, we will configure our E2E instance to read files from a specific pattern and set supportFolder and supportFile to false because we don't want to include any support files or folders for this demo.

In the next section, we will create our first e2e test file and test out our login functionalities using E2E testing.

Creating the test file

To create the test file, open or create a Login.spec.js file inside the e2e test folder and add the following code:

```
/* eslint-disable no-under */
const loginFunction = () => {
  cy.visit(`${Cypress.env("baseUrl")}/login`);
};
describe("Login tests", () => {
  beforeEach(() => {
    loginFunction();
```

```
    cy.wait(5000);
  });
  it("Should show an success message if the email address
      and password is valid", () => {
    cy.get("#passwordField").type(`${Cypress.env("password")}`);
    cy.get("#emailField").type(`${Cypress.env("email")}`);
    cy.get("#loginForm").then(() => {
      cy.get("#submitButton").click();
      cy.wait(1000);
      cy.get("#loggedIn").should("be.visible");
    });
  });
  it("Should show an error message if the email address and
      password is not valid", () => {
    cy.get("#emailField").type("test@test.com");
    cy.get("#passwordField").type("test");
    cy.get("#loginForm").then(() => {
      cy.get("#submitButton").click();
      cy.wait(1000);
      cy.get("#failed").should("be.visible");
    });
  });
});
});
```

Code walk-through:

Let's walk through the code together and understand the nitty-gritty of it:

- *Step 1*: Loading the login page inside `beforeEach`:

  First, we create the function to visit our login page using the Cypress `visit()` method that will be executed inside the `beforeEach` hook:

  ```
  const loginFunction = () => {
    cy.visit(`${Cypress.env("baseUrl")}/login`);
  };
  ```

- *Step 2*: Creating the `beforeEach` block:

  Inside the `beforeEach` block, we execute `loginFunction` to open the login page for every test case:

  ```
  describe("Login tests", () => {
    beforeEach(() => {
      afterLoginFunction();
      cy.wait(5000);
    });
  ```

- *Step 3*: Writing each test case:

  Lastly, we start writing each test case and defining what we expect to test. The following is an example of using E2E testing to submit a button in our login form to mimic how a user will interact with the login form:

  ```
  it("Should show a success message if the email address
     and password is valid", () => {
    cy.get("#passwordField").type(`${Cypress.
  env("password")}`);
    cy.get("#emailField").type(`${Cypress.env("email")}`);
    cy.get("#loginForm").then(() => {
      cy.get("#submitButton").click();
      cy.wait(1000);
      cy.get("#loggedIn").should("be.visible");
    });
  });
  ```

After writing all your test cases, you can execute your test. Before you run your test, make sure that your development server is up and running.

Running the test

To run your test, type the following command into your root Terminal session:

```bash
```bash
yarn test:e2e
yarn test:e2e:ci
```
```

The first command will visually show you how your users will interact with your application using a headless browser, while the last command will only show you the result of your test such as unit tests.

After successfully running the test, you should be greeted with green passes for your test, as shown in the following screenshot:

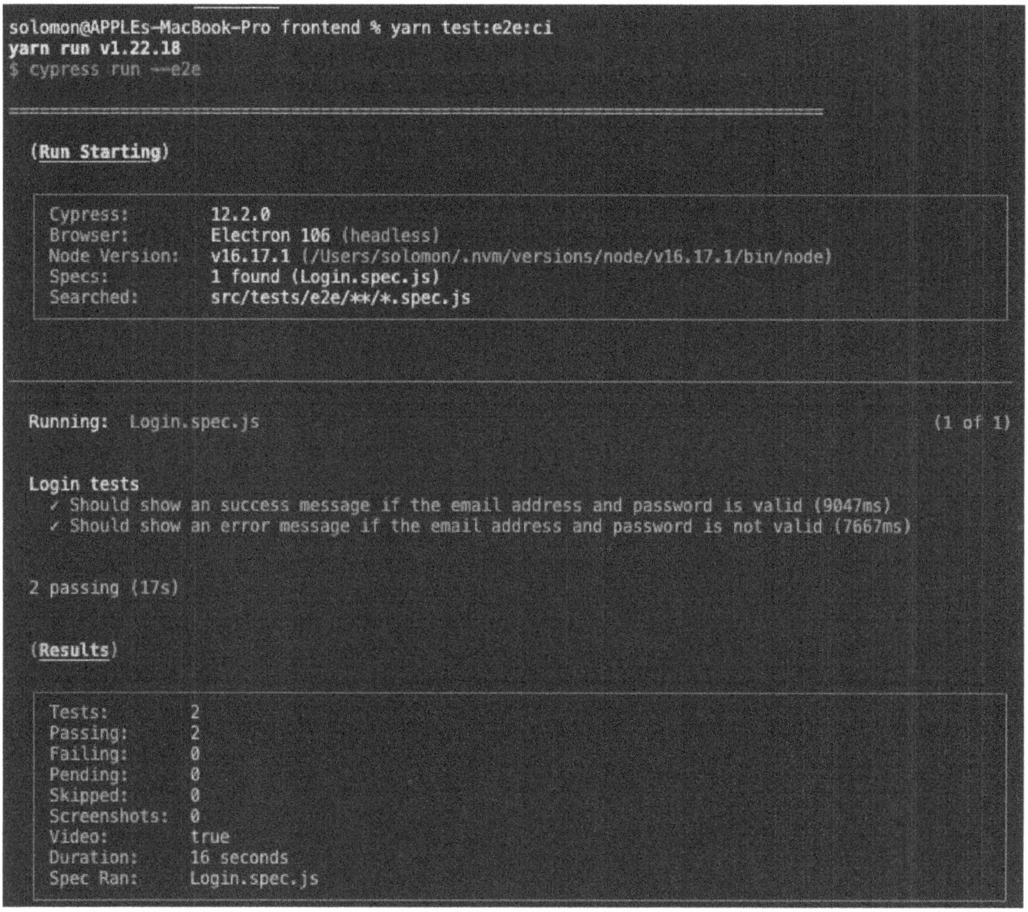

```
solomon@APPLEs-MacBook-Pro frontend % yarn test:e2e:ci
yarn run v1.22.18
$ cypress run --e2e

========================================================================

  (Run Starting)

  ┌────────────────────────────────────────────────────────────────────┐
  │ Cypress:       12.2.0                                               │
  │ Browser:       Electron 106 (headless)                             │
  │ Node Version:  v16.17.1 (/Users/solomon/.nvm/versions/node/v16.17.1/bin/node) │
  │ Specs:         1 found (Login.spec.js)                             │
  │ Searched:      src/tests/e2e/**/*.spec.js                          │
  └────────────────────────────────────────────────────────────────────┘


  Running:  Login.spec.js                                         (1 of 1)


  Login tests
    ✓ Should show an success message if the email address and password is valid (9047ms)
    ✓ Should show an error message if the email address and password is not valid (7667ms)


  2 passing (17s)


  (Results)

  ┌────────────────────────────────────────────────────────────────────┐
  │ Tests:        2                                                     │
  │ Passing:      2                                                     │
  │ Failing:      0                                                     │
  │ Pending:      0                                                     │
  │ Skipped:      0                                                     │
  │ Screenshots:  0                                                     │
  │ Video:        true                                                  │
  │ Duration:     16 seconds                                           │
  │ Spec Ran:     Login.spec.js                                        │
  └────────────────────────────────────────────────────────────────────┘
```

Figure 11.6 – A screenshot of the E2E test result

Summary

This chapter explored everything related to E2E testing. We learned about E2E testing tools such as Cypress, Karma, and Selenium and used Cypress to effectively test an enterprise project.

Additionally, we explored the benefits, importance, and best practices in writing and executing effective E2E testing strategies. We also learned how to write E2E test cases.

In the next chapter, we will learn how to deploy Vue.js 3 projects to the AWS cloud. We will learn about the best practices for deploying to AWS. In addition, we will learn how enterprise companies deploy their enterprise Vue applications.

Additionally, we will learn about and explore different deployment options and master best practices in deploying your Vue.js 3 project to various cloud providers. We will learn how to deploy to AWS and Azure.

Part 5: Deploying Enterprise-ready Vue.js 3

In this part, you will learn and explore different deployment options and master best practices to deploy your Vue.js 3 project to various cloud providers. You will learn how to deploy to AWS and Azure.

We will explore Nuxt.JS to build and deliver enterprise-ready, server-side rendering Vue.js 3 web applications. We will also explore Gridsome to build and deploy high-, client-side rendering Vue.js 3 applications.

In this part, we will cover the following chapters:

12

Deploying Enterprise-Ready Vue.js 3

In the previous chapter, we explored everything related to **end-to-end** (**e2e**) testing. We learned in depth how to perform e2e testing on a Vue.js 3 component and pages. In addition, we also learned about end-to-end testing tools such as Cypress and Puppeteer, and how to use them to test an enterprise project effectively.

In this chapter, we will learn how to deploy Vue.js 3 projects to the AWS cloud. We will learn the best practices for deploying to AWS. In addition, we will learn how enterprise companies deploy their enterprise Vue.js 3 applications.

Additionally, we will learn about and explore different deployment options and best practices to deploy your Vue.js 3 project to various cloud providers. We will learn how to deploy the app to AWS and Azure.

We will cover the following key topics in this chapter:

- Introduction to CI/CD
- Overview of CI/CD
- What is a deployment pipeline?
- Overview of GitHub Actions
- Deploying to AWS

Technical requirements

To get started, I recommend reading through *Chapter 11, Industry Standard End-to-End Testing*, where we explored the concept of e2e testing and what to test from an array of components and methods available. We will rely heavily on the knowledge of that chapter in this chapter to learn about Vue.js 3 enterprise unit testing.

All the code files of this chapter can be found at `https://github.com/PacktPublishing/Architecting-Vue.js-3-Enterprise-Ready-Web-Applications/tree/chapter-12`.

Introduction to CI/CD

Developing an enterprise-level application is easy, but constantly deploying newly developed changes, bug fixes, or features to your users is a daunting process, especially if done frequently, and especially for enterprise-ready applications. In addition, as your application, teams, and deployment infrastructure grows in complexity, continuously releasing and deploying new changes, features, and products to customers can be a complicated process.

To solve the complicated process of developing, testing, and releasing software quickly and consistently, three related but distinct strategies have been created by developers and organizations to manage and automate these processes.

In the next section, we will explore these three pillars, called CI/CD, and explain each of these strategies and how they relate to each other. Most importantly, we will explore how to build and incorporate these strategies into our enterprise application life cycle so that it can transform our software development and release practices.

Overview of CI/CD

CI/CD stands for **continuous integration/continuous delivery**. It is a strategy that allows enterprise teams to ship software faster and more efficiently. It enables a streamlined approach for getting products to the market more quickly than ever before, allowing for a steady stream of code to be released into production and providing a steady stream of new features and bug fixes through the most efficient means of delivery.

A CI/CD pipeline is written to automate the software delivery process from the development stage to the production environment. It builds, tests, and safely deploys a new version of an application.

The main advantage of automated pipelines is that it removes the manual errors that can be detected during deployment and provides standardized feedback loops to developers for faster product iterations.

CI/CD is a combination of different strategies and pillars that come together to create a strong pipeline for delivering enterprise software; we will explore these strategies in detail in this section.

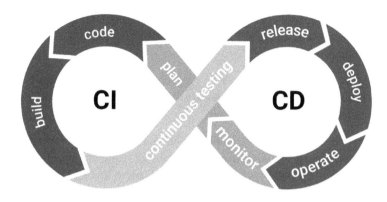

Figure 12.1 – CI/CD

Continuous integration

CI is a process that allows developers in a team to frequently integrate their code into a shared repository. These developers can write their code in isolation and integrate it using a continuous integration process. This practice encourages each developer to build in isolation and integrate code with the shared repository multiple times throughout the day.

When code is integrated early in the development cycle, developers can discover conflicts at the boundaries between new and existing code early. This process minimizes the cost of integration by making it an early consideration.

By implementing a proper continuous integration strategy, development teams can reduce integration costs and respond to defects early.

For an enterprise team to succeed in robust, automated, and fast integration, deployment, and delivery of enterprise software, the culture of frequent iteration and responsiveness to build issues must be cultivated.

Continuous delivery

CD is an extension of continuous integration that is aimed at streamlining the software delivery process and allowing teams to deploy their code to production with ease and assurance. It seeks to reduce the difficulty of the deployment or release process by automating the steps necessary to deploy a build, thus enabling code to be released securely at any time.

Additionally, continuous delivery is a process that allows for the automated transfer of finished code to various settings, such as testing and development. It provides a reliable and automated method for the code to be sent to these areas.

In addition, continuous delivery encompasses the provisioning and deployment of infrastructure, which can be done manually and involve multiple steps. This type of delivery usually automates these processes with the involvement of the entire team.

Continuous delivery relies on a deployment pipeline to automate the process of running increasingly comprehensive test suites against a build, with each stage being a sequential step. If the build fails the test, the team is notified, but if it passes, it is automatically advanced to the next stage.

It is essential for enterprise software teams to implement continuous delivery, as it automates the process between committing code to the repository and determining whether to deploy well-tested, functional builds to the production environment. This step helps ensure that the quality and accuracy of the code are automated, while the ultimate decision of what to release is left up to the engineering team.

Continuous deployment

Continuous deployment is an extension of continuous delivery that deploys each build that passes the full test cycle without manual intervention. This is beneficial, as manual deployment can cause delays and irregular deployment. A continuous deployment system will deploy any build that has gone through the deployment pipeline that was set up during the continuous delivery stage.

In addition, deploying your code automatically doesn't mean that new features cannot be activated or deactivated conditionally. In fact, continuous deployment can be configured to only deploy a specific feature to a subset of users or be activated conditionally at a later time.

The debate surrounding continuous deployment is often focused on the safety of automated deployment and whether the risk it poses is worth the reward. Nevertheless, it can also be advantageous for organizations, as they can receive constant feedback on new features and quickly detect any errors before too much time and energy is wasted.

We have explored the concept of CI/CD and how to automate deploying and releasing enterprise projects. In the next section, we will explore the deployment pipeline and how to create an enterprise-ready deployment pipeline for the enterprise Vue.js 3 application.

What is a deployment pipeline?

The deployment pipeline streamlines the deployment and delivery of your enterprise application. It compiles the code, executes all the tests, and securely deploys a new version of the application.

Automating your deployment and delivery processes using deployment pipelines removes manual errors, provides standardized feedback loops to developers, and enables fast product iterations.

Furthermore, when building enterprise products, your organizational structure and development team and pattern will determine the strategies used to create your deployment pipeline, as it can differ from project to project.

However, there are different strategies already used in enterprise projects that can be adopted and modified if necessary.

In deployment pipelines, there are required stages (or elements) that make up a CI/CD pipeline. In the next section, we are going to explore these elements and learn how to set up our deployment pipeline for our demo enterprise project.

Elements of a deployment pipeline

A deployment pipeline is composed of executable instructions that any developer must follow in order to release a new version of a software product.

The beauty of an automated deployment pipeline is that it replaces the manual process of carrying out the exact specification laid out for the deployment and delivery of enterprise software by automating the process.

The following figure shows the typical software release stages in most enterprise software:

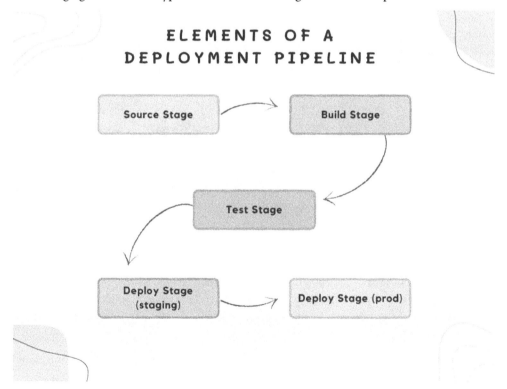

Figure 12.2 – Elements of a deployment pipeline

These stages can be performed manually, provided each step is followed accordingly. However, the downside is enormous, as you can see here:

- **Time-consuming**: Manual deployment can take a long time to complete, especially if there are multiple components that need to be deployed
- **Error-prone**: Manual deployment is prone to human errors, which can lead to costly mistakes and downtime
- **Lack of scalability**: Manual deployment is not easily scalable, as it requires manual intervention for each component that needs to be deployed

To avoid this, an automated deployment pipeline has been created to carry out the stages and alert the responsible developer of any errors, or to send notifications through email, Slack, and so on. Additionally, the pipeline can also notify the whole team when a successful deployment to production has been completed.

Now, let's examine each of the stages to understand what goes in. This will aid us in understanding how to develop a good deployment pipeline for our enterprise Vue.js 3 application.

Source stage

At the source stage, a pipeline is typically initiated by a source code repository. Whenever there is a change in the code, it notifies the CI/CD process to execute the related pipeline. Additionally, other common triggers include user-initiated workflows and automated scheduled workflows.

Build stage

In the build stage, we combine the source code and all its dependencies to build a runnable instance of the project that is intended to ship to the users. At this stage, the software is compiled or bundled together with its dependencies.

The build phase attempts to package the project to make it deployable. If the build stage encounters any problems, this is a sign of an underlying issue with the project's setup or configuration and should be taken care of right away.

Test stage

Once the build stage is finished successfully, the next step is to conduct the test stage. This stage involves running automated tests to make sure the code is accurate and the project is functioning correctly. This stage serves as a safeguard to ensure that any bugs that can be easily reproduced are not sent to the end users or passed through the pipeline.

At this stage, all the test cases written by developers (including, but not limited to, unit tests, integration tests, e2e testing, etc.) are all tested and checked to make sure they all pass before allowing the current build to proceed to the deployment stage.

The testing stage is critical for identifying any issues with the code that the developer may have overlooked. This feedback is important to the developer as it is provided while the problem is still fresh in their mind. If any failures occur during the test stage, they can reveal problems in the code.

Deploy stage

Prior to this stage, the pipeline has created a functioning version of the new code or modifications that have passed all the predetermined tests and is now ready to deploy it.

Generally, there are multiple deployment environments that have been established for the development team, such as a "staging" environment for the product team, a "development" environment for the development team, and a "production" environment for the end users.

Depending on the team, organization, and model chosen, various deployment environments can be established. Teams that have adopted the Agile model of development, which is based on tests and real-time observation, often deploy to the staging environment for further manual testing and examination before pushing out accepted modifications to production for the end users.

Overview of a deployment pipeline

In this section, we are going to explore a practical example of a deployment pipeline. Pipelines are the reflection of the complexity of a project. Therefore, configuring a pipeline that runs on every code change will save a team many pains and repetitive tasks in the future.

The following figure shows a clear example of a deployment pipeline and the different jobs that need to be performed.

The `source` stage is triggered when changes are pushed to a specific branch that the CI/CD is activated on and it moves to the build stage where it compiles that code using a compiler, if any, or uses a `docker build` process to build the project's image.

Next, the testing stage runs all the necessary and activated test cycles, such as unit tests, integration tests, and end-to-end tests.

After successful testing, the pipeline moves to the deploy stage, where the code is deployed to a live staging environment for further testing before finally deploying to the production environment.

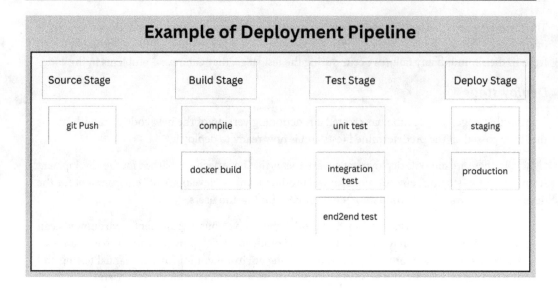

Figure 12.3 – Example of a deployment pipeline

In the preceding figure, we have explored the overview of a deployment pipeline, the different stages, and what goes under the hood of each different stage. In this section, we have explored deployment pipelines and the different stages that are involved in them. Next, we will discuss how to deploy our Pinterest demo to the AWS cloud using GitHub Actions.

Overview of GitHub Actions

In the world of CI/CD, there are numerous tools have been created to automate the process of building, testing, and deploying projects. GitHub Actions happens to be one of those tools and has greatly gained popularity.

GitHub Actions is a CI/CD platform that allows developers to automate the process of building, testing, and running deployment pipelines.

GitHub Actions became popular because it is directly integrated into GitHub and can be configured to create workflows that build and test every pull request to your repository or deploy merged pull requests to production.

There are tons of concepts to learn about GitHub Actions: the different terminologies, concepts, benefits, and advantages of using GitHub Actions over other CI/CD platforms. You can learn all this from the official documentation at `https://docs.github.com/en/actions/learn-github-actions/understanding-github-actions`.

Nevertheless, we will show you how to create a deployment pipeline for the Pinterest demo project we developed in previous chapters.

Deploying an enterprise project is tedious and requires lots of checks to make sure that frequent bugs do not surface in production.

There are many factors to check before deploying an enterprise project, from linting, formatting, and styling, to testing. The list is endless and sometimes depends on your team and how the development workflow is set up.

In the next section, we are going to explore the different stages or checks the project needs to pass before deploying to production.

Jobs in the deployment pipeline

The jobs in the deployment pipeline vary from project to project and from team to team. In the following subsection, we look at some of the important jobs you can build into your deployment pipeline to completely check your project before deploying to production.

Linting (Eslint, Stylelint, Prettier)

Linting is a process in which a linter program reviews the source code of a particular programming language or code base to detect any potential issues such as errors, bugs, stylistic errors, and suspicious constructs. This is beneficial in recognizing both common and uncommon mistakes that can be made when coding. Furthermore, linting will go through your source code to identify any formatting discrepancies, check for compliance with coding standards and conventions, and pinpoint potential logical errors in your program.

Also, linting helps with developer experience in a team by creating a consistent code base throughout the development team.

We are going to set up linting in our pipeline to enforce consistency between the style guide, formatting, and naming conventions in our deployment pipeline, like so:

```
lint:
  runs-on: ubuntu-latest
  steps:
    - uses: actions/checkout@v3
    - run: |
        yarn
        yarn lint
```

Lighthouse budget checks

Lighthouse is an open source, automated tool for improving the quality of web pages. This tool allows you to run tests against web pages (public or requiring authentication). It helps developers audit web pages for performance, accessibility, SEO, and more.

In addition, you can automate this process by adding it to your deployment pipeline to test for performance of your web page before deploying it to the users. This process allows enterprise-ready application developers to automate the process of testing the performance of the application in real time.

The action allows us to set numerous options, including the following:

- Testing against multiple paths
- Providing a budget path
- The number of runs (how many times the CI should audit an URL)

We are going to set up the Lighthouse bot (`https://github.com/ebidel/lighthousebot`) with GitHub Actions to audit our deployment and User Experience (UX) integrity.

Here is a snippet of the job setup for Lighthouse:

```
lighthouse:
    runs-on: ubuntu-latest
    needs: deploy
    steps:
      - uses: actions/checkout@v2
      - name: Run Lighthouse on urls and validate with
            lighthouserc
        uses: treosh/lighthouse-ci-action@v7
        with:
          urls: |
            ${{ needs.deploy.outputs.preview-url }}
          budgetPath: ./budget.json
          runs: 3
```

The preceding code is used to add the Lighthouse plugin to the deployment pipeline, and the plugin uses the `actions/checkout` plugin to access the repository workspace in other to access the `budget.json` file, which contains the task that Lighthouse should perform. This Lighthouse GitHub Action is extremely beneficial for websites that depend on Google search traffic. If not addressed early on, it is very common for the bundle sizes to become larger as a website is developed, resulting in a lower Lighthouse score. This action allows us to monitor any discrepancies with each commit.

Automated software testing

Software testing is a vital factor for determining the status of your enterprise application and making sure it conforms to the project requirements. As explored in the previous chapters, we have developed three major types of software testing and have practiced how to create better testing suites for our Pinterest application demo.

Therefore, we are going to set up three jobs to run the entire software testing suites for our demo application. In our demo, the software testing suite comprises the following:

- Unit testing

- Integration (component) testing

- E2e testing

The job will run each of these tests and respond accordingly. If the test fails, it will pause the deployment and report the problem to the development team via Slack notifications or emails. Otherwise, if the test passes, it will continue to the next stages.

Here is the snippet of all the test setups:

```
unit_test:
  runs-on: ubuntu-latest
  steps:
    - uses: actions/checkout@v3
    - run: |
        yarn
        yarn test:unit


component_test:
  runs-on: ubuntu-latest
  needs: unit_test
  steps:
    - uses: actions/checkout@v3
    - run: |
        yarn
        yarn test:component

e2e_test:
  runs-on: ubuntu-latest
  needs: component_test
  steps:
    - uses: actions/checkout@v3
    - run: |
        yarn
        yarn test:e2e
```

The script sets up the testing stage, which contains scripts to run different testing cycles such as unit testing, integration testing, and e2e testing. In each of the pipeline jobs, we use `actions/checkout` to check out the workspace repository, and next, we run the `yarn` command to install all the packages before proceeding to run the `test` command.

Netlify deployment for staging

Netlify is a comprehensive platform that enables you to integrate your preferred tools and APIs to construct the fastest websites, stores, and applications for the composable web. It allows you to utilize any frontend framework to construct, preview, and deploy to the worldwide network from Git.

You can deploy your enterprise application to several environments, such as development, staging, and production, depending on your team's setup.

GitHub Actions allows you to create several workflows for deploying to these different environments. Within each environment, you can set up different jobs to be performed. For instance, you might not want to check for Lighthouse performance again since it was already tested when deploying to staging environments.

Here is a snippet to set up the job to deploy to Netlify:

```
deploy:
   runs-on: ubuntu-latest
   needs: e2e_test
   steps:
     - uses: actions/checkout@v2
     - name: Deploy to Netlify
       uses: nwtgck/actions-netlify@v1.2
       id: deploy-to-netlify
       with:
         publish-dir: './dist'
         production-branch: master
         github-token: ${{ secrets.GITHUB_TOKEN }}
         deploy-message: "Deploy from GitHub Actions"
         enable-pull-request-comment: false
         enable-commit-comment: true
         overwrites-pull-request-comment: true
       env:
         NETLIFY_AUTH_TOKEN: ${{ secrets.NETLIFY_AUTH_TOKEN
                             }}
         NETLIFY_SITE_ID: ${{ secrets.NETLIFY_SITE_ID }}
```

```
        timeout-minutes: 1
    outputs:
      preview-url:
        ${{ steps.deploy-to-netlify.outputs.deploy-url }}
```

The preceding script uses the Netlify GitHub Action plugin to deploy the Vue.js 3 application to Netlify. It requires a Netlify token and secrets (which are added in the Secrets section of our GitHub repository) and finally, it provides the preview URL after deployment.

In the next section, we are going to create a complete deployment pipeline with GitHub Actions to set up a staging application for more manual testing before pushing it to the master branch, which will trigger the production deployment pipeline.

Creating the deployment pipeline with GitHub Actions

To create a deployment pipeline with GitHub Actions, we need to create configuration files for each pipeline configuration environment.

Follow the steps mentioned next to create your first deployment pipeline for your staging environment using GitHub Actions.

Open the Pinterest demo application or clone it from the official repository for this chapter to see a complete setup of the GitHub Actions.

If you're following along, create a new file called staging.yml inside the .github/workflows folder.

It's important to note that the name of the folders must be exactly the same for GitHub Actions to pick the configuration up when pushing to your repository.

Pipeline for the staging environment

Open the staging.yml file and add the following scripts to create a deployment pipeline for the staging environment:

```
on:
pull_request:
  branches:
    - chapter-12
jobs:
lint:
  runs-on: ubuntu-latest
  steps:
    - uses: actions/checkout@v3
```

```
      - run: |
          yarn
          yarn lint
unit_test:
  runs-on: ubuntu-latest
  steps:
    - uses: actions/checkout@v3
    - run: |
        yarn
        yarn test:unit


component_test:
  runs-on: ubuntu-latest
  needs: unit_test
  steps:
    - uses: actions/checkout@v3
    - run: |
        yarn
        yarn test:component


e2e_test:
  runs-on: ubuntu-latest
  needs: component_test
  steps:
    - uses: actions/checkout@v3
    - run: |
        yarn
        yarn test:e2e


deploy:
  runs-on: ubuntu-latest
  needs: e2e_test
  steps:
    - uses: actions/checkout@v2
    - name: Deploy to Netlify
      uses: nwtgck/actions-netlify@v1.2
```

```
      id: deploy-to-netlify
      with:
        publish-dir: './dist'
        production-branch: master
        github-token: ${{ secrets.GITHUB_TOKEN }}
        deploy-message: "Deploy from GitHub Actions"
        enable-pull-request-comment: false
        enable-commit-comment: true
        overwrites-pull-request-comment: true
      env:
        NETLIFY_AUTH_TOKEN: ${{ secrets.NETLIFY_AUTH_TOKEN }}
        NETLIFY_SITE_ID: ${{ secrets.NETLIFY_SITE_ID }}
      timeout-minutes: 1
    outputs:
      preview-url:
        ${{ steps.deploy-to-netlify.outputs.deploy-url }}


  lighthouse:
    runs-on: ubuntu-latest
    needs: deploy
    steps:
      - uses: actions/checkout@v2
      - name: Run Lighthouse on urls and validate with
              lighthouserc
        uses: treosh/lighthouse-ci-action@v7
        with:
          urls: |
            ${{ needs.deploy.outputs.preview-url }}
          budgetPath: ./budget.json
          runs: 3
```

In this section, we discussed in detail how to automate the process of deploying your enterprise applications. We learned about deployment pipelines and how to create one with GitHub Actions. In the next section, we will learn how to deploy our app to **AWS** (**Amazon Web Services**) production by automating the process using deployment pipelines.

Deploying to AWS

In this section, we are going to implement continuous deployment for the Vue.js 3 application with GitHub Actions and AWS App Runner.

This process can be triggered manually after thoroughly checking the staging application to make sure it satisfies all requirements before pushing it to production. However, it can also be automated to happen immediately after the staging is completed.

In this demo, we are going to create the deployment pipeline for deploying to the AWS production server using AWS App Runner and also automate the process at once.

> **Important note**
>
> It's advisable to trigger the deployment process manually, which gives room to manually check all the requirements on the staging environment before deploying a new release to production.

To deploy to AWS, you will need an AWS account and an AWS IAM account with proper permissions. In this section, we explored how to create pipelines and deploy our project to AWS. In the next section, we will continue deploying our project using Docker and the Dockerfile we created in the previous chapters.

Using Docker

In *Chapter 7, Dockerizing a Vue 3 App* we discussed the nitty-gritty involved in Dockerizing your Vue.js 3 project. In addition, we learned about the best practices and industry standards to Dockerize and deploy an enterprise Vue.js 3 web application.

We will use the Dockerfile we created for this project so that we can run it on AWS infrastructure as a containerized application.

Update the Dockerfile we created before with the following code snippet:

```
# Use the official Node.js 14 Alpine image from https://hub.
docker.com/_/node.
# Using an image with specific version tags allows
deterministic builds.
FROM node:fermium-alpine3.14 AS builder


# Create and change to the app directory.
```

```
WORKDIR /app

# Copy important root files to the builder image.
COPY package*.json ./

# Install production dependencies.
RUN npm install

# Copy the Vue 3 source to the container image.
COPY . .

# build app for production with minification
RUN npm run build

# Production stage
FROM nginx:stable-alpine as production-stage

# Copy the Vue 3 source to the container image.
COPY --from=builder /app/dist /usr/share/nginx/html

VOLUME /app/node_modules

EXPOSE 80

# Run the Vue service on container startup.
CMD ["nginx", "-g", "daemon off;"]
```

This is the same Dockerfile we used in the previous chapter to Dockerize our project. You can refer back to the chapter to learn more about Dockerizing your Vue.js 3 application.

The base image will be nginx:stable-alpine and the application will be listening on port 80. For step-by-step Dockerizing guidelines, please refer to the official documentation from Vue.js at https://v2.vuejs.org/v2/cookbook/dockerize-vuejs-app.html?redirect=true.

You can test the application container using the following Docker Compose command since we have already defined the `docker-compose.yml` file in *Chapter 7, Dockerizing a Vue 3 App*:

```
docker build -t vue-paper-dashboard .
docker run -dt --name vue-paper-dashboard -p 8080:80 vue-paper-
dashboard:latest
```

After running the application container successfully, you should be able to access the dashboard via the same address as the `npm run dev` command. Next, let's provision AWS resources.

Provisioning AWS resources

We will use GitHub Actions to deploy our Vue.js 3 application to AWS continuously, so we need to create an IAM account and a user-managed role on AWS, which will be used in the next steps.

Creating an IAM account

This IAM account will be used by GitHub Actions agents. Access the console at `https://us-east-1.console.aws.amazon.com/iamv2/home#/home` and create an IAM account and a user-managed role on AWS.

Add user   ① ② ③ ④ ⑤

Set user details

You can add multiple users at once with the same access type and permissions. Learn more

User name* | vue-pinterest-demo |

➕ Add another user

Select AWS access type

Select how these users will primarily access AWS. If you choose only programmatic access, it does NOT prevent users from accessing the console using an assumed role. Access keys and autogenerated passwords are provided in the last step. Learn more

Select AWS credential type* ☑ **Access key - Programmatic access**
Enables an **access key ID** and **secret access key** for the AWS API, CLI, SDK, and other development tools.

☐ **Password - AWS Management Console access**
Enables a **password** that allows users to sign-in to the AWS Management Console.

* Required   Cancel   **Next: Permissions**

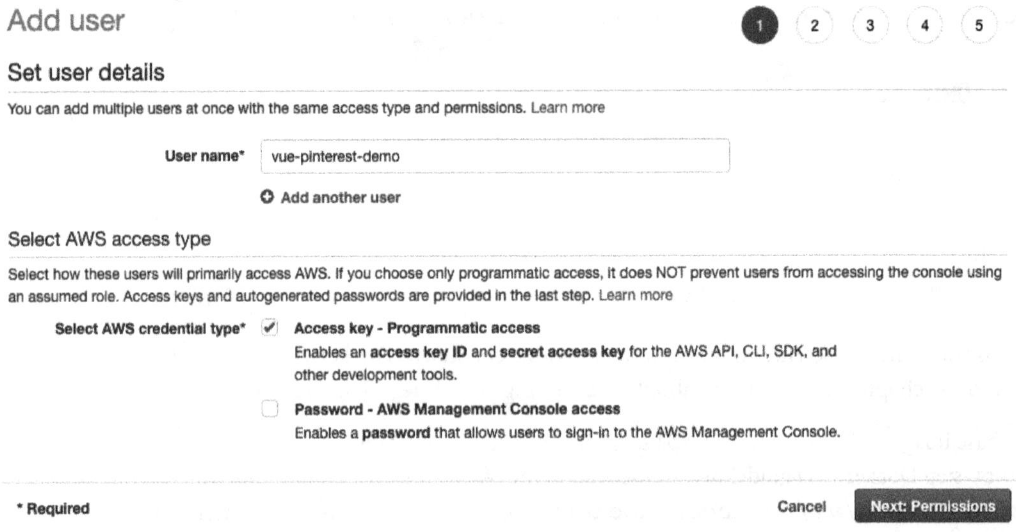

Figure 12.4 – Creating a new user in IAM for GitHub Actions

Click on the **Next: Permissions** option and click the **Create User** button. Lastly, click on **Download. csv** to download the credential for the new user and save it somewhere—we will need to use it soon.

Creating a role for the IAM user

In this demonstration, we will be making a new role called `app-runner-service-role` and attaching the `AWSAppRunnerServicePolicyForECRAccess` policy. This role will be used by AWS App Runner services to give them access to Elastic Container Register (ECR) in order to manage our Docker image.

To create a service role, follow these steps:

1. Click on the **Role** menu.

2. Click on the **Create Role** button, select the **Custom Trusted Policy** option, and enter the following JSON:

```
{
    "Version": "2012-10-17",
    "Statement": [
        {
            "Effect": "Allow",
            "Principal": {
                "Service":
                    "build.apprunner.amazonaws.com"
            },
            "Action": "sts:AssumeRole"
        }
    ]
}
```

This code snippet is a JSON file that is used to create a custom trusted policy for deploying to Amazon **Elastic Container Service (ECS)**.

After successfully creating `app-runner-service-role`, as shown in the following figure, make sure to copy and note the Amazon Resource Name (ARN), as it will be used later.

Figure 12.5 – Creating app-runner-service-role

In this section, we worked through the step-by-step process of creating `app-runner-service-role` and the Amazon IAM permission for ECS deployment. In the next section, we will be creating a policy for the IAM user.

Creating a policy for the IAM user

Navigate to the `github-vue-pinterest-demo` IAM permission and attach the following inline policy, which will grant permission to GitHub Actions (via the IAM role) to work with ECR and App Runner:

```
{
    "Version": "2012-10-17",
    "Statement": [
        {
            "Sid": "VisualEditor0",
            "Effect": "Allow",
            "Action": "apprunner:*",
            "Resource": "*"
        },
        {
```

```json
                "Sid": "VisualEditor1",
                "Effect": "Allow",
                "Action": [
                    "iam:PassRole",
                    "iam:CreateServiceLinkedRole"
                ],
                "Resource": "*"
        },
        {
                "Sid": "VisualEditor2",
                "Effect": "Allow",
                "Action": "sts:AssumeRole",
                "Resource": "{ENTER_YOUR_SERVICE_ROLE_ARN_HERE}"
        },
        {
                "Sid": "VisualEditor3",
                "Effect": "Allow",
                "Action": [
                    "ecr:GetDownloadUrlForLayer",
                    "ecr:BatchGetImage",
                    "ecr:BatchCheckLayerAvailability",
                    "ecr:PutImage",
                    "ecr:InitiateLayerUpload",
                    "ecr:UploadLayerPart",
                    "ecr:CompleteLayerUpload",
                    "ecr:GetAuthorizationToken"
                ],
                "Resource": "*"
        }
    ]
}
```

By updating the IAM policy to be more specific (i.e., ARN-specific instead of wildcard), the security concerns associated with the preceding JSON can be addressed by creating a policy and attaching it to the IAM user.

Creating an ECR private repository

We're almost there; one final step is to create a private repository on ECR to manage our Docker images. Add a repository name of your choice into the box provided, and click on the **Create** button, leaving the remaining options as their defaults.

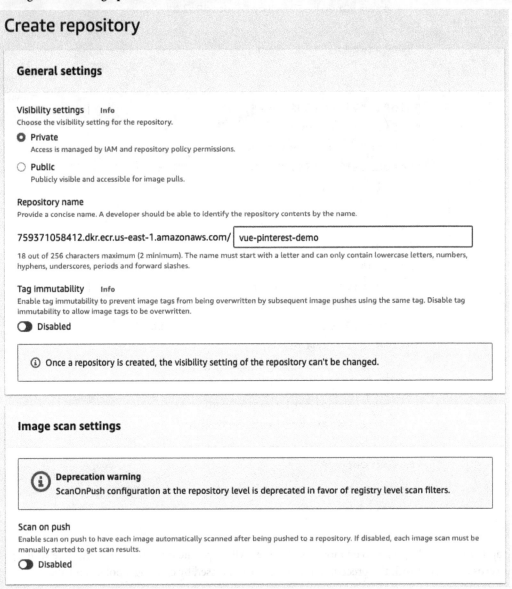

Figure 12.6 – Creating a private repository named vue-pinterest-demo

After creating your ECR instance, head over to your GitHub repository and add all the secrets and environment variables needed to deploy your application.

In this section, we created the ECR instance and added our secrets to our GitHub repository, along with all the environment variables needed. In the next section, we will look at how to work with GitHub Actions to automate the deployment process.

Working with GitHub Actions

In this section, we will be working with GitHub Actions and automating the process of deploying your application to Amazon ECR. We will start by adding the Amazon secrets to our GitHub repository. Follow these steps to add your secrets:

1. Go to **Settings** | **Secrets** | **Actions** in your GitHub repository and add all the necessary secret variables.

2. Open the `new_user_credentials.csv` file you downloaded when you created the IAM user.

3. Copy the values for **AWS_ACCESS_KEY_ID** and **AWS_SECRET_ACCESS_KEY** and paste them into your GitHub Secrets as your environment variable.

4. Additionally, you can use `us-east-1 for AWS_REGION` and your ARN of `app-runner-service-role` for ROLE_ARN.

After adding your credentials successfully, in the next section, we will create a pipeline for deploying the enterprise project to AWS App Runner using ECR to manage our Docker images.

Pipeline for the production environment

Open the `production.yml` file and add the following scripts to create a deployment pipeline for the production environment:

```
name: PRODUCTION - Deploy container to AWS App Runner
on:
 push:
   branches:
     - master
 workflow_dispatch: # Allow manual invocation of the
                    # workflow
env:
 ENVIRONMENT_NAME: production
 ECR_REPOSITORY_NAME: vue-pinterest-demo
jobs:
```

```yaml
deploy:
  runs-on: ubuntu-latest

  steps:
    - name: Checkout
      uses: actions/checkout@v2
      with:
        persist-credentials: false

    - name: Configure AWS credentials
      id: aws-credentials
      uses: aws-actions/configure-aws-credentials@v1
      with:
        aws-access-key-id: ${{ secrets.AWS_ACCESS_KEY_ID }}
        aws-secret-access-key:
          ${{ secrets.AWS_SECRET_ACCESS_KEY }}
        aws-region: ${{ secrets.AWS_REGION }}

    - name: Login to Amazon ECR
      id: ecr-login
      uses: aws-actions/amazon-ecr-login@v1

    - name: Build, tag, and push image to Amazon ECR
      id: build-image
      env:
        ECR_REGISTRY: ${{ steps.ecr-login.outputs.registry }}
        ECR_REPOSITORY: ${{ env.ECR_REPOSITORY_NAME }}
        IMAGE_TAG: ${{ github.sha }}
      run: |
        docker build -t
        $ECR_REGISTRY/$ECR_REPOSITORY:$IMAGE_TAG .
        docker push
        $ECR_REGISTRY/$ECR_REPOSITORY:$IMAGE_TAG
        echo "::set-output name=
          image::$ECR_REGISTRY/$ECR_REPOSITORY:$IMAGE_TAG"
```

```
- name: Deploy to App Runner
  id: deploy-app
  uses: awslabs/amazon-app-runner-deploy@main
  with:
    service: erp-app-${{ env.ENVIRONMENT_NAME }}
    image: ${{ steps.build-image.outputs.image }}
    access-role-arn: ${{ secrets.ROLE_ARN }}
    region: ${{ secrets.AWS_REGION }}
    cpu : 1
    memory : 2
    port: 80
    wait-for-service-stability: false

- name: App Runner output
  run: echo "App runner output
    ${{ steps.deploy-app.outputs.service-id }}"
```

If everything is successful, navigate to the App Runner service console; there will be a service with the name you specified. You can click on the default domain name to preview your application or set up a custom domain name.

Figure 12.7 – Preview of the Pinterest demo application

Summary

In this chapter, we learned how to deploy Vue.js 3 projects to the AWS cloud and some of the best practices for deploying to AWS. We explored continuous integration and continuous delivery by exploring deployment pipelines, showing the different deployment staging, and how to configure each of them to perform specific jobs. We also examined how each stage performs and how to deploy to a staging environment.

Additionally, we explored different deployment options and how to deploy using Docker with Amazon ECR. We learned practically how to create an account and set up Amazon ECR with Docker and finally, we implemented automated deployment using CI/CD, Docker, Amazon ECR, and GitHub Actions.

In the next chapter, we will explore the definitive guide to Nuxt.js. You will learn the nitty-gritty of Nuxt.js and how to build and deliver enterprise SSR projects with Vue.js 3. We will explore the definitive guide to Gridsome, and you will learn the nitty-gritty of Gridsome and how to build and deliver enterprise CSR projects with Vue.js 3.

13
Advanced Vue.js Frameworks

In the previous chapter, we explored how to deploy Vue.js 3 projects to AWS Cloud. We learned about the best practices for deploying to AWS. In addition, we learned how enterprise companies deploy their enterprise Vue applications.

This chapter explores a definitive guide to Nuxt.js. We will learn about the nitty-gritty of Nuxt.js and how to build and deliver enterprise SSR projects with Vue.js 3. Additionally, we will explore a definitive guide to Gridsome where you will learn about the nitty-gritty of Gridsome and how to build and deliver enterprise CSR projects with Vue.js 3.

In this chapter, we will cover the following key topics:

- Introduction to Vue frameworks
- Top Vue frameworks
- Understanding Nuxt.js and how it works
- Benefits of Nuxt.js
- Creating a Nuxt.js app
- What is Gridsome?
- What is Gridsome used for?
- How does Gridsome work?
- Benefits of Gridsome
- Creating a Gridsome app

Technical requirements

To get started with this chapter, you should read through *Chapter 12, Deploying Enterprise-Ready Vue.js 3*, where we learned how to deploy Vue.js 3 projects to the AWS cloud and some of the best practices for deploying to AWS. Additionally, we explored different deployment options and mastered best practices in deploying our Vue.js 3 project to AWS. We will rely heavily on the knowledge from that chapter in this chapter to learn about more advanced Vue frameworks.

Introduction to Vue frameworks

A single framework cannot solve all the problems of frontend engineering, as it will become bloated and very heavy to load. Vue.js is not exempt; there are some issues that are not completely out-of-the-box with Vue.js. Also, it will require harder configuration and can lead to a waste of development time to implement some features into Vue.js right out of the box.

The pressing needs of developers make Vue.js the framework of all frameworks. In recent years, after the release of the Vue.js framework, we have noticed a good number of other frameworks that have been created out of Vue.js.

These frameworks offer different options to meet development needs such as **Server-Side Rendering (SSR)**, **Static Site Generators (SSGs)**, **Progressive Web Apps (PWAs)**, and more.

Frameworks can be divided into four distinct groups based on their purpose and capabilities. These include the following:

- **Vue.js UI frameworks**: These provide tools to create modern, responsive websites
- **Mobile frameworks**: These help to construct hybrid mobile web apps
- **Static site frameworks**: These generate static websites
- **SSR frameworks**: These are used to create SSR applications

In this section, we are going to explore the top Vue.js frameworks, and later in this chapter, we will explore the top two frameworks in more detail.

Top Vue frameworks

There are various Vue.js frameworks available for developers to utilize when creating universal Vue applications. Let's explore some of them in the following subsections.

Vue UI frameworks

Creating an attractive and user-friendly UI is a key part of frontend development. If the design of the interface is not appealing or easy to use, it will be hard to keep a consistent user base.

When designing a UI for a large enterprise product, it is beneficial to utilize a Vue UI framework that offers pre-made Vue components and elements to create an attractive frontend.

The top UI component frameworks for Vue are as follows:

- Bootstrap Vue
- Vuetify
- Quasar Framework
- Vue Material

You can compare the popularity of each of the frameworks using this *npm trends* link: `https://npmtrends.com/bootstrap-vue-vs-quasar-vs-vue-material-vs-vuetify`.

Mobile frameworks

Over the years, Vue.js has gained popularity in hybrid and robust web app development including hybrid mobile development according to Monocubed (`https://www.monocubed.com/blog/why-vuejs-gaining-popularity/`).

However, both hybrid and native mobile development can also be achieved with Vue.js using some of the mobile frameworks listed here in conjunction with other mobile development libraries and frameworks:

- Vue Native
- Vux
- Mint UI

You can compare the popularity of each of the frameworks using this *npm trends* link: `https://npmtrends.com/mint-ui-vs-vue-native-core-vs-vux`.

Static site frameworks

A SSG utilizes templates and raw data to create static HTML pages. One of the advantages of this is that the website loads in the same manner each time, and the content does not vary dynamically, meaning that the web pages do not need to be coded separately.

The following is a list of Vue frameworks used to generate static websites:

- Vue Press
- Gridsome
- Nuxt.js

Moreover, it's important to note that Nuxt.js can be used to generate static websites. However, that is not the main focus of Nuxt.js.

You can compare the popularity of each of the frameworks using this *npm trends* link: `https://npmtrends.com/gridsome-vs-nuxt-vs-vuepress`.

SSR frameworks

According to the official documentation (`https://vuejs.org/guide/scaling-up/ssr.html#why-ssr`), SSR applications have better and faster time-to-content conversion rates, better SEO, and the same unified language and declarative component-oriented metal model for developing your entire application.

A Vye.js application rendered server-side allows your application codes to run on both the server and client side as opposed to SSGs, which will only run on the client side.

The following is a list of Vue frameworks used to implement SSRs:

- Nuxt.js
- Quasar
- Vite SSR

You can compare the popularity of each of the frameworks using this *npm trends* link: `https://npmtrends.com/nuxt-vs-quasar-vs-vite-ssr`.

In this section, we explored the top Vue.js frameworks and their different categories. In the next section, we will take a deep-dive into Nuxt.js and see how it works.

Understanding Nuxt.js and how it works

Nuxt.js is an open source, Vue.js-based framework that provides developers with the tools to create frontend projects with confidence. It is designed to make web development simpler and more powerful, and it also offers server-side rendering capabilities to help developers manage complex configurations for asynchronous data, middleware, and routing:

Figure 13.1 – The official Nuxt.js logo

Vue.js applications can be organized using a well-known architecture, which can be used to create either basic or complex applications. Furthermore, this structure can help to improve the development of Vue.js applications.

In this section, we will learn about the different use cases of Nuxt.js and why you should consider switching to Nuxt.js.

Uses of Nuxt.js

With Nuxt.js, you're limitless in terms of the type of applications you can build, and Nuxt.js has been used to develop high-performing and SEO-focused websites. In the following subsections, we will look at the most popular types of websites you can use to build with Nuxt.js.

Statically generated pages

Statically generated pages are websites that do not require any outside data sources, as the content is already included in the HTML. Nuxt.js can be used to create statically generated pages, such as portfolios, demo sites, or tutorial pages.

SPAs

A **single-page application** (**SPA**) is a type of frontend development that retrieves data from an external source and displays it on a user's device. It is not surprising that many popular JavaScript frameworks such as React.js, Vue.js, and Angular are all SPA frameworks.

The HTML 5 history Application Programming Interface (API) and the location hash are utilized to create SPA routing systems. This capability allows developers to alter a website's URL without needing to reload the entire page.

Universal applications

This part of Nuxt.js is my favorite because almost all the applications I have developed with Nuxt.js have been universal applications.

A universal application is a technique that utilizes SSR to obtain client-side data on the server before completely displaying the page on the client's web browser.

SSR is built into Nuxt.js right out of the box, and it solves the tedious configurations that are involved in activating and enabling SSR in Vue.js.

Nuxt.js can be used to resolve the existing SSR issue in Vue.js, which is beneficial for SEO and can even be extended to create a universal application that allows for a single code base to be used for both the frontend and the backend of a monolithic application.

These were just some of the categories of applications you can use to build with Nuxt.js. In the next section, we are going to explore how Nuxt.js actually works.

How does Nuxt.js work?

Depending on your settings, Nuxt.js can operate in two different ways. If you enable SSR or use the universal mode, it will function in the same manner as a server-side framework. This means that every time a user visits your website, the requests are processed on the server, and a server is required to render and deliver the page.

However, if client-side rendering is enabled or universal mode has not been activated, the content and the pages are rendered in the browser using JavaScript. This approach has the fastest load time and performs well in terms of speed and page performance.

The Nuxt.js lifecycle gives a high-level overview of the different parts of the framework, their order of execution, and how they work together. Also, it describes what happens after the build phase, where your application is bundled, chunked, and minified.

There are three main actions and methods used in Nuxt.js depending on whether you enabled SSR or not:

- The `nuxtServerInit` action is the initial hook that is executed on the server side if a Vuex store is enabled. It is used to fill the store and is only called if the store has been enabled. Additionally, this hook can be used to dispatch other actions in the Vuex store on the server.

- `validate()` is a function that validates the dynamic parameters of a page component. It is called before rendering the page components.

- `AsyncData` and `Fetch` are functions utilized to acquire data and display it on the server side (`AsyncData`) or to obtain data and fill the store before rendering the page (`Fetch`).

Here is a quick summary of how your requests are processed when you visit a Nuxt.js website. When Nuxt.js receives an initial page visit, it calls out to the `nuxtServerInit` action to update the store or dispatch necessary actions if your store is enabled; otherwise, Nuxt.js will ignore `nuxtServerInit` and move to the next stage.

Next, Nuxt.js will look up your `nuxt.config.js` file for any global middleware and execute it accordingly. After the execution, it will move to the layout pages and check for any middleware for execution, and lastly, it will execute the page's middleware including the page children.

After executing the middleware in order, it will check the routes and use the `validate()` function to run validations against the params, queries, and more.

The `asyncData` method is then employed to acquire and display data on the server side if it had been enabled previously. Afterward, the `fetch` method is used to fill Vuex on the client side.

At this point, the page should have all the required data to be displayed a proper web page. The following diagram of a flowchart illustrates all the steps it takes to render a single page:

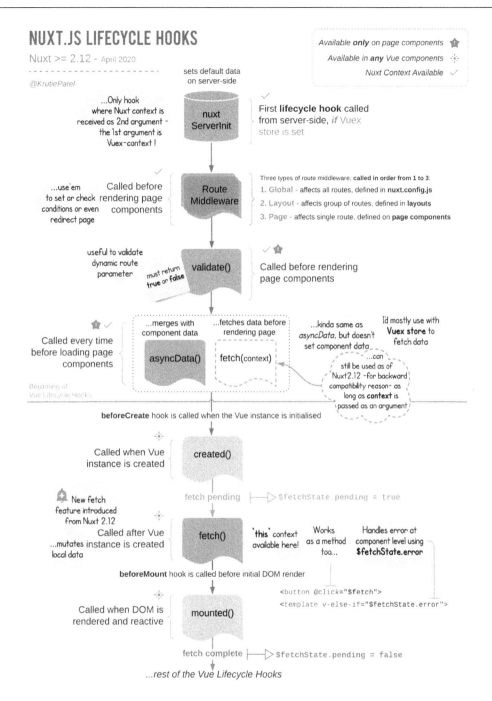

Figure 13.2 – An overview of Nuxt.js lifecycle hooks (source: https://nuxtjs.org/docs/concepts/nuxt-lifecycle/)

The official Nuxt.js lifecycle (`https://nuxtjs.org/docs/concepts/nuxt-lifecycle`) page gives a more detailed overview of the behind-the-scenes of how Nuxt.js renders and processes your pages whether they are enabled server-side or client-side.

Having gone through the inner workings of Nuxt.js, it should now be simple to comprehend. In the following section, let's investigate the advantages of using Nuxt.js for your upcoming project.

Benefits of Nuxt.js

The benefits of Nuxt.js cannot be underestimated; you can easily spot a few of them with the introduction of SSR and project structuring for enterprise-level projects.

However, in the following subsections, we will understand some of the benefits of the Nuxt.js framework and why it's becoming very popular for building SSR-enabled projects with Vue.

Creating universal apps easily

With Nuxt.js, you can create SSR applications very easily without needing to go through the painful route of configuring Vue to support SSR. The SSR feature is already built into Nuxt.js and is very easy to use.

Nuxt.js exposes two important properties called `isServer` and `isClient` to determine the state of the framework at runtime. It can be useful when checking whether your component should render on the server side or the client side.

Statically rendering your Vue apps with universal benefits

Statically generated websites are actively gaining popularity with different frameworks developed to focus only on them. However, you can easily generate a static website with Nuxt.js without installing any additional frameworks or tools.

You can quickly create a static version of your website, complete with HTML and routes, by using the `nuxt generate` command.

Nuxt.js enables the creation of a powerful universal application that does not require a server to utilize the SSR feature, similar to building a statically generated website.

Automatic code-splitting

Frontend development that focuses on speed and performance has become a fundamental part of enterprise software, and Nuxt.js stands out for its exceptional performance due to its code-splitting feature.

This feature allows each route to be given its own JavaScript file that only contains the code necessary to run that route. This approach to building applications helps to reduce the amount of code that needs to be loaded in order to render a single page, thus decreasing loading times.

Webpack's built-in configuration enables code splitting when creating static web pages for your website.

ES6/7 compilation

ES6 and 7 are enabled by default in Nuxt.js because Webpack and Babel are prebuilt into it for translating and compiling the latest version of JavaScript to the versions that older browsers can execute.

Babel is set up to take all the `.vue` files and ES6 code within the script tags and convert them into JavaScript, which is compatible with all browsers. This functionality eliminates the need to manually set up and configure browser compatibility from the beginning.

In the next section, we will look at how to create our first Nuxt.js application and the practical approach for developing enterprise-ready applications with Nuxt.js.

Creating a Nuxt.js app

This section will introduce you to a practical approach to developing applications with Nuxt.js. Before we delve in, let's explore some of the few critical concepts when it comes to developing enterprise-ready applications with Nuxt.js.

Creating a Nuxt application

You can easily create a Nuxt.js application in different ways, but the recommended way is to use any of the following commands:

```
Yarn create nuxt-app <project-name>
Or
npm init nuxt-app <project-name>
Or
npx create-nuxt-app <project-name>
```

Next, move into the created project folder and serve your newly created Nuxt.js project with the following command:

```
cd <project-name>
npm run dev
Or
yarn dev
```

It's important that you replace `<project-name>` with an actual project name.

Now that we have our new Nuxt.js project generated for us, let's explore the different folders and files that come with the project.

Understanding the Nuxt.js folder structure

When you create a new project using any of the preceding commands mentioned, it can feel quite daunting due to the number of folders and files it comes with. In this section, we will take a look at some of the key folders and files that are part of the Nuxt.js project. Additionally, some of these files and folders are essential and must remain unchanged without any extra configuration. The following figure shows the folder structure of Nuxt.js:

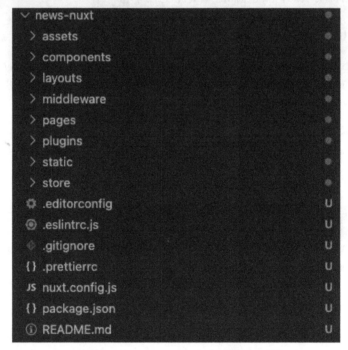

Figure 13.3 – A screenshot of the Nuxt.js folder structure

Let's go over this folder structure in the following subsections.

.nuxt

When you start your development server, the `.nuxt` folder will be created automatically and will not be visible. This folder is also known as the build directory and includes generated files and artifacts that are used to serve your project during development.

assets

The `assets` directory holds all the raw materials such as pictures, CSS, SASS documents, fonts, and more. Webpack will compile any file that is included in this folder while the page is being created.

components

This folder is analogous to the `components` folders in Vue. It is the repository of all your Vue components. Components are the files that make up the various components of your pages and can be reused and imported into any page, layout, or component.

layouts

The `layouts` folder is a great place for organizing the different page layouts of your application. It can be used to differentiate between the page structure of the dashboard and the page structure for users who are not logged in. This helps to keep the different parts of the application organized.

You can create different structures to correspond to different structures of your applications such as different sidebars, menus, headers, footers, and more. You can achieve all these separations with Nuxt.js layouts.

middleware

Middleware can be defined as custom functions that are triggered before or after a page or set of pages (layout) is rendered. These middleware functions can be stored in the `middleware` folder in Nuxt.js.

Middleware is important and handy when creating membership-only or authentication-enabled applications. You can use it to restrict users from accessing certain authenticated pages.

plugins

The `plugins` directory is where all the JavaScript code you want to run before initializing the root Vue application is stored. This is the place in which to add Vue plugins and inject functions or constants.

You will use this folder a lot to include different Vue plugins that have not been included in Nuxt.js as a module.

It works by creating a JavaScript file in the `plugins` folder, using the `Vue.use()` function to add the plugin to the Vue instance, and lastly, adding the file to the `plugins` array in the `nuxt.config.js` file.

static

The `static` directory is a special one, housing all the static files of your application that are unlikely to be altered or that will be displayed without any further processing by Nuxt.js or Webpack.

Any files located in the `static` folder will be provided by Nuxt.js and can be accessed through the root URL of the project. This includes items such as `favicon`, `robot.txt`, and more.

store

The store folder holds all of your Vuex store documents, and it is automatically divided into modules. The Vuex store is included in the package, but it has to be enabled by creating an index.js file in the store folder before it can be used.

Nuxt.js is designed to help with the development of enterprise-level applications, and it comes pre-equipped with Vuex for state management. This makes it easier to create and manage applications of this scale.

pages

The pages folder is very important as it is the bedrock of the Nuxt.js routing system. Therefore, it cannot be renamed without updating the Nuxt.js configuration. Nuxt.js automatically reads all the .vue files inside the pages directory and creates a corresponding route for the page.

The pages directory holds all the views and routes for your application, and each page component is a regular Vue component that Nuxt.js automatically transforms into routes by adding special attributes and functions to make the development of your application smooth and straightforward.

In the next section, we will explore how Nuxt.js automatically converts .vue files in the pages folder into routes.

Nuxt.js pages and routing system

Nuxt.js simplifies the routing process by allowing users to create directories and files in the pages folder, which will then automatically generate a router file based on the structure of the directory.

For example, if you have a posts.vue file in the directory, it will automatically be converted into a route, and you can then access the route in your browser to view the content of the Posts page.

This routing system enables you to establish three distinct routes simply by creating files and folders. Let's take a closer look at these route types.

We will explore the different types of routing that are supported by Nuxt.js and see how each of the routing types is used within Nuxt.js.

Basic routing

Routing is a process by which requests are routed or directed to the code that handles them. These requests can come in the form of URLs and are redirected to the appropriate handler. It can be a simple process, as no extra configuration is needed for it to function. Examples of this are /about, /contact, /posts, and more. To set up a basic routing system, the pages directory should be organized in the following manner:

```
pages/
  -| about.vue
  -| contact.vue
  -| posts.vue
```

Nuxt will generate a router file automatically similar to the following:

```
router: {
  routes: [
    {
      name: 'posts',
      path: '/posts',
      component: 'pages/posts.vue'
    },
    {
      name: 'about',
      path: '/about',
      component: 'pages/about'
    },
    {
      name: 'contact',
      path: '/contact',
      component: 'pages/contact'
    },
  ]
}
```

The preceding code snippet is automatically generated by Nuxt.js and is not available for editing because everything is properly routed according to your folder structure in the pages directory.

Nested routing

Nested routes are routes that are embedded within a parent route. This type of routing is used to create multiple levels of routing that are more detailed.

With Nuxt.js, you can easily create nested routes by creating a parent folder and placing all the route files within that folder. Take a look at the following folder structure:

```
pages/
 --| dashboard/
 -----| portfolios.vue
 -----| settings.vue
 --| dashboard.vue
 --| posts.vue
 --| contact.vue
 --| index.vue
```

In the preceding code, we created a new file and folder with the same name as the dashboard in the directory structure shown previously. Afterward, we placed the `portfolios.vue` and `settings.vue` files as sub-items in the dashboard folder.

This straightforward organization of folders will create a router with routes that will look like the following:

```
router: {
  routes: [
    {
      name: 'index',
      path: '/',
      component: 'pages/index.vue'
    },
    {
      name: 'posts',
      path: '/posts',
      component: 'pages/posts'
    },
    {
      name: 'contact',
      path: '/contact',
```

```
        component: 'pages/contact'
    },
    {
      name: 'dashboard',
      path: '/dashboard',
      component: 'pages/dashboard.vue',
      children: [
        {
          name: 'dashboard-portfolios',
          path: '/dashboard/portfolios',
          component: 'pages/dashboard/portfolios.vue'
        },
        {
          name: 'dashboard-settings',
          path: '/dashboard/settings',
          component: 'pages/dashboard/settings.vue'
        }
      ]
    }
  ]
}
```

In Vue.js, nested routes are created manually and registered inside an `index.js` router file, which can easily becomes complicated when creating many routes for an enterprise application, but with Nuxt.js, it is made very simple and easy to create files and nested folders.

Dynamic routing

Dynamic routes can be generated with undefined route names either due to an API call or because you don't want to keep creating the same page. These routes are generated from a variable, such as a name or ID, which is obtained from the data within the application.

In order to make a route dynamic, you must add an underscore at the end of the `.vue` file or directory name. You can name the file or directory whatever you want, but an underscore must be included in order for it to be dynamic.

For instance, if you define a `_slug.vue` file in the `pages` directory, you can access the value using the `params.slug` object.

Using a dynamic route is advantageous when constructing a blog application; for instance, when the ID or slug of the post that the user is going to select to read is unknown. However, with a dynamic route, it is possible to obtain the ID/slug of the post and display the appropriate post with the ID/slug.

Using an example, we'll demonstrate how to create a dynamic route:

```
pages/
--| posts/
-----| _slug.vue
-----| index.vue
--| services.vue
--| contact.vue
--| index.vue
```

Here, we add an underscore to `slug` in order to create a dynamic route for the page, as well as a parent route with a string parameter and its respective child routes. This page structure will generate a router with the following routes in the file:

```
{
  name: 'index',
  path: '/,
  component: 'pages/index.vue'
},
{
  name: 'contact',
  path: '/contact',
  component: 'pages/contact.vue'
},
{
  name: 'services',
  path: '/services',
  component: 'pages/services.vue'
},
{
  name: 'posts',
  path: '/posts',
```

```
      component: 'pages/posts/index.vue',
      children: [
        {
          name: 'posts-slug,
          path: '/posts/:slug,
          component: 'pages/posts/_slug.vue'
        }
      ]
    }
  ]
}
```

Now that we've explored the different routing systems that come built into the Nuxt.js framework, you have a solid knowledge of how Nuxt.js works and can start building your enterprise-ready universal application with it.

In the next section, we will explore Gridsome and learn about the nitty-gritty of Gridsome and how to build and deliver enterprise CSR projects with Vue.js 3.

What is Gridsome?

Gridsome is a powerful static website generator. It is powered by Vue.js to build statically generated websites and apps that are fast by default. It is also a Jamstack framework for building websites and applications that delivers better performance, higher security, and lower cost of scaling.

Gridsome is focused on implementing the Jamstack approach to build fast and secure sites and applications by pre-rendering files and serving them directly from a CDN – thereby increasing the speed of your application and removing the requirement to manage or run web servers.

Jamstack is an architectural approach that decouples the web experience layer from data and business logic, improving flexibility, scalability, performance, and maintainability.

What is Gridsome used for?

At the moment, Gridsome does not support SSR but focuses on creating faster websites and applications. In the following subsections, we will look at the most popular types of websites you can build with Gridsome.

Statically generated pages

These are the types of websites that do not require any external data sources – the content is already embedded into the HTML. You can use Gridsome to create statically generated pages such as portfolios, demo websites, or tutorial pages with different data sources and higher performance.

SPAs

A frontend development approach that utilizes dynamic data from an external API and displays it on the client side is known as creating an SPA. It is not unexpected that the majority of JavaScript frameworks, such as React.js, Vue.js, and Angular, are all SPA frameworks.

The HTML 5 history API and the location hash are utilized to create SPA routing systems. This capability allows developers to alter a website's URL without needing to reload the entire page.

In the next section, we will explore how Gridsome works and how can use it to create a statically rendered website.

How does Gridsome work?

Gridsome is a Jamstack framework; therefore, it uses modern web development architecture based on client-side JavaScript, reusable APIs, and prebuilt markup.

It works by generating static SEO-friendly HTML markup that is converted into a dynamic DOM once loaded in the browser. This simple feature allows Gridsome to be a go-to Jamstack framework for building both static and dynamic websites.

Internally, Gridsome builds one `.html` file and one `.json` file for every page that you create and loads the `.json` files after the first page load to prefetch and load data for the next pages. Additionally, it also builds a `.js` bundle for each page to take advantage of code splitting.

Additionally, the source plugins can obtain data from either local files or external **APIs** and store it in a local database. A unified GraphQL data layer allows you to access only the required data from the database and use it in your Vue components.

The following diagram explains the inner workings of Gridsome and how data is passed and processed until it gets to your Vue component:

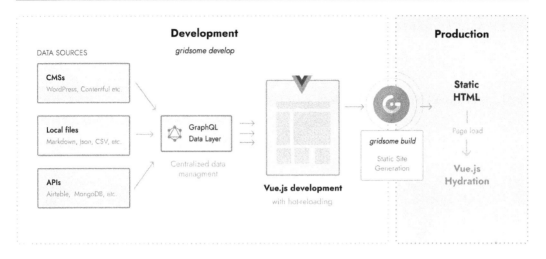

Figure 13.4 – An overview of how Gridsome works (source: https://gridsome.org/docs/how-it-works/)

There are two ways to run Gridsome:

- `gridsome develop`: This command starts a local development server and watches for changes
- `gridsome build`: This command generates production-ready static files

You can learn more about how each command works and how they generate these static pages from the official documentation at `https://gridsome.org/docs/how-it-works/`.

Next, let's look at some benefits of using Gridsome.

Benefits of Gridsome

The benefits of using Gridsome are enormous and depend on the type of website and application intended. For statically generated websites, Gridsome proves to focus on speed by default and has good project structures for building enterprise-ready static websites and applications.

The following subsections describe some of the benefits of using Gridsome.

Serverless and statically generated

Gridsome uses the Jamstack approach of building websites thats, which provides better performance and increased security, and reduces costs and complexity in your development stack. Gridsome generates static pages and websites using the Jamstack philosophy where the final product is a folder with static HTML files that can be deployed anywhere.

Easy to install and use

Gridsome is very easy to install and simple to start using. It comes with a CLI, which is a command-line tool that helps you to create Gridsome projects effortlessly.

You can install Gridsome by running the following command:

```
npm install -global @gridsome/cli
```

Once the CLI has been installed, you can use it to create as many Gridsome projects as you want in the future.

Organized project structure

One of the challenges of enterprise projects is the project structure, as discussed in previous chapters. Gridsome solves this problem by helping you structure your enterprise project with a predictive project structure.

We will go over the important files and folders that make up the Gridsome project in the *Understanding the Gridsome folder structure* section.

Automatic routing

Automatic routing is a very important feature in the frontend development industry, starting with Nuxt where routes are automatically generated as you add files and folders to the pages folder.

Gridsome also makes it super easy to create routing with the automatic routing feature. Routes are generated automatically whenever there's a file or new folder in the src/pages folder. This is similar to how Nuxt works, as discussed earlier.

Code splitting/pre-fetching

With the integration of code splitting and pre-fetching in Gridsome, navigation in a Gridsome website becomes super fast because any link you click on has already been prefetched before you clicked on it.

Additionally, the code-splitting feature helps in increasing the performance and loading speed of a Gridsome website because it allows the user to only load only the JavaScript that is needed to only load the requested page and load the others on demand.

Markdown file support

Using markdown in Gridsome is the easiest way to automate your content management. You can create content in the form of blog posts, articles, or anything described in its own .md (markdown extension) file. These markdown files will be grouped and consumed by Gridsome to generate individual HTML files.

Now that we have seen the benefits of using Gridsome, let's see how to create a Gridsome app in the next section.

Creating a Gridsome app

This section will introduce you to a practical approach to developing applications with Gridsome. Before we delve in, let's explore a few critical concepts in developing enterprise-ready applications with Gridsome.

Creating a Gridsome application

You can easily create a Gridsome application in different ways, but the recommended way is to use any of the following commands:

```
Yarn global add @gridsome/cli
Or
npm install - -global @gridsome/cli
```

Next, move into the created a Gridsome project and serve your newly created Gridsome project with the following command:

```
gridsome create <project-name>
cd <project-name>
gridsome develop
```

It's important that you replace `<project-name>` with an actual project name.

Now that we have our newly Gridsome project generated for us, let's explore the different folders and files that come with the project.

Understanding the Gridsome folder structure

When you scaffold a new project using any of the preceding commands, it comes with a lot of overwhelming folders and files. But in this section, we will explore some of the important folders and files within the newly created Gridsome project.

Furthermore, some of these files and folders are vital and require that some of the folder names and filenames remain unchanged without additional configuration. Here is what the Gridsome folder structure looks like:

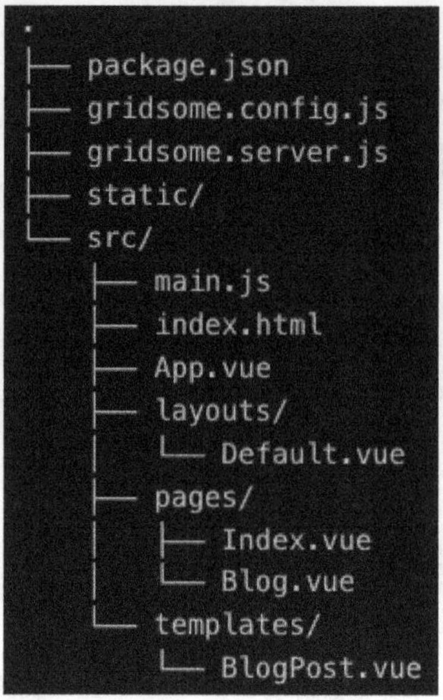

```
.
├── package.json
├── gridsome.config.js
├── gridsome.server.js
├── static/
└── src/
    ├── main.js
    ├── index.html
    ├── App.vue
    ├── layouts/
    │   └── Default.vue
    ├── pages/
    │   ├── Index.vue
    │   └── Blog.vue
    └── templates/
        └── BlogPost.vue
```

Figure 13.5 – The Gridsome folder structure

Let's go over some of the important folders in the following subsections.

pages

The pages folder is one of the most important folders in Gridsome as it is responsible for automatic routing and works exactly the same as in Nuxt except each page will be generated statically and have its own index.html file with markup.

There are two options for creating pages in Gridsome:

- **File-based pages**:

    When creating your pages with single file components (a single .vue file), then you should use the filesystem. Any single file component found in the src/pages directory will automatically be converted into its own route or URL. The file location is used to generate the URL, and you can see it in the following example:

    ```
    pages/
      —| about.vue
      —| contact.vue
      —| posts.vue
    ```

The preceding page structure will be converted into the following:

```
`pages/about.vue` becomes `/about`
`pages/contact.vue` becomes `/contact`
`pages/posts.vue` becomes `/posts`
```

Next, let's explore the second option of creating pages in Gridsome called programmatic pages.

- **Programmatic pages**:

The `createPages` hook, located in the `gridsome.server.js` file, can be used to generate programmatic pages. This is useful if you need to manually create pages from an external API without using Gridsome's built-in GraphQL data layer.

You can programmatically create a page by implementing the `createPages` hook, as shown in the following code block:

```
module.exports = function (api) {
  api.createPages(({ createPage }) => {
    createPage({
      path: '/my-new-page',
      component: './src/templates/MyNewPage.vue'
    })
  })
}
```

You can also create dynamic pages using the same `createPages` hook as shown earlier.

Templates

Gridsome uses templates to display nodes or single pages of collections. When you create a template file, Gridsome will try to locate a file with the same name as the collection if not specified in the template config. Often, templates are mapped to collections for displaying information.

Here is an example of displaying a post title from a GraphQL query using templates:

```
<!-- src/templates/Post.vue -->
<template>
  <Layout>
    <h1 v-html="$page.post.title" />
  </Layout>
</template>
```

```
<page-query>
query ($id: ID!) {
  post(id: $id) {
    title
  }
}
</page-query>
```

Templates are very important in Gridsome as they are a way to present data pages in their own URLs. You can learn more advanced use cases of templates from the documentation at `https://gridsome.org/docs/templates/`.

Layouts

Layouts are Vue components that are used inside pages and templates to wrap the content. You can use a layout to create different structures for your website. It works exactly like layouts in Nuxt.js.

Usually, layouts are used as follows in pages:

```
<template>
  <Layout>
    <h1>About us</h1>
  </Layout>
</template>

<script>
import Layout from '~/layouts/Default.vue'

export default {
  components: {
    Layout
  }
}
</script>
```

In Gridsome, layouts files are global and do not need to be imported before you start using them. There are more important files and folders that come with the Gridsome project, and as your project grows, you will quickly discover that you have added even more files and folders.

However, the previously mentioned folders are the important folders and their names should not be changed as you add more files and folders to your project.

Summary

In this chapter, we explored every important detail about Nuxt.js. You learned about the nitty-gritty of Nuxt.js and how to build and deliver enterprise SSR projects with Vue.js 3. Additionally, we explored Gridsome, the super fast Jamstack framework for building statically generated websites.

We also covered the benefits of using Nuxt and Gridsome. Then, we saw how to create an app using both of these frameworks. Finally, we explored the folder structures of Nuxt and Gridsome.

Index

www.packtpub.com

Subscribe to our online digital library for full access to over 7,000 books and videos, as well as industry leading tools to help you plan your personal development and advance your career. For more information, please visit our website.

Why subscribe?

- Spend less time learning and more time coding with practical eBooks and Videos from over 4,000 industry professionals

- Improve your learning with Skill Plans built especially for you

- Get a free eBook or video every month

- Fully searchable for easy access to vital information

- Copy and paste, print, and bookmark content

Did you know that Packt offers eBook versions of every book published, with PDF and ePub files available? You can upgrade to the eBook version at packtpub.com and as a print book customer, you are entitled to a discount on the eBook copy. Get in touch with us at customercare@packtpub.com for more details.

At www.packtpub.com, you can also read a collection of free technical articles, sign up for a range of free newsletters, and receive exclusive discounts and offers on Packt books and eBooks.

Other Books You May Enjoy

If you enjoyed this book, you may be interested in these other books by Packt:

ASP.NET Core and Vue.js

Devlin Basilan Duldulao

ISBN: 9781800206694

- Discover CQRS and mediator pattern in the ASP.NET Core 5 Web API

- Use Serilog, MediatR, FluentValidation, and Redis in ASP.NET

- Explore common Vue.js packages such as Vuelidate, Vuetify, and Vuex

- Manage complex app states using the Vuex state management library

- Write integration tests in ASP.NET Core using xUnit and FluentAssertions

- Deploy your app to Microsoft Azure using the new GitHub Actions for continuous integration and continuous deployment (CI/CD)

Vue.js 3 By Example

John Au-Yeung

ISBN: 9781838826345

- Get to grips with Vue architecture, components, props, directives, mixins, and other advanced features
- Understand the Vue 3 template system and use directives
- Use third-party libraries such as Vue Router for routing and Vuex for state management
- Create GraphQL APIs to power your Vue 3 web apps
- Build cross-platform Vue 3 apps with Electron and Ionic
- Make your Vue 3 apps more captivating with PrimeVue
- Build real-time communication apps with Vue 3 as the frontend and Laravel

Packt is searching for authors like you

If you're interested in becoming an author for Packt, please visit `authors.packtpub.com` and apply today. We have worked with thousands of developers and tech professionals, just like you, to help them share their insight with the global tech community. You can make a general application, apply for a specific hot topic that we are recruiting an author for, or submit your own idea.

Hi!

I am Solomon Eseme, author of *Architecting Vue.js 3 Enterprise-Ready Web Applications*. I really hope you enjoyed reading this book and found it useful for increasing your productivity and efficiency in Vue.js.

It would really help me (and other potential readers!) if you could leave a review on Amazon sharing your thoughts on *Architecting Vue.js 3 Enterprise-Ready Web Applications*.

Go to the link below or scan the QR code to leave your review:

`https://packt.link/r/1801073902`

Your review will help us to understand what's worked well in this book, and what could be improved upon for future editions, so it really is appreciated.

Best Wishes,

Solomon Eseme

Download a free PDF copy of this book

Thanks for purchasing this book!

Do you like to read on the go but are unable to carry your print books everywhere? Is your eBook purchase not compatible with the device of your choice?

Don't worry, now with every Packt book you get a DRM-free PDF version of that book at no cost.

Read anywhere, any place, on any device. Search, copy, and paste code from your favorite technical books directly into your application.

The perks don't stop there, you can get exclusive access to discounts, newsletters, and great free content in your inbox daily

Follow these simple steps to get the benefits:

1. Scan the QR code or visit the link below

https://packt.link/free-ebook/9781801073905

2. Submit your proof of purchase

3. That's it! We'll send your free PDF and other benefits to your email directly